The Florida Gardener's
ANSWER BOOK

Felice Dickson

The Florida Gardener's
ANSWER BOOK

Banyan Books
Miami, Florida

Designed by Bernard Lipsky

Manufactured in the United States of America
by Miami Book Manufacturing, Inc., Miami, Florida

Library of Congress Cataloging in Publication Data
Dickson, Felice
    The Florida gardener's answer book.

    (Tropical gardening series ; 2)
    Bibliography: p.
    1.   Gardening—Florida.   2.   Tropical plants—Miscel-
lanea. I. Title.
SB453.2.F6D5   635'.09759   76-3534
ISBN 0-916224-02-3

NOTE: Due to continuing review by the federal
Environmental Protection Agency all pesticide
recommendations in this book should be considered current
as of publication date only. Check with your
County Cooperative Extension Service for changes.

Contents

Foreword

Gardeners inspired and fascinated by tropical plants can't stop asking and answering questions as they become intrigued by and learn more about everything from ylang-ylangs (they're flowering trees, in case you're wondering) to camu-camus (they're fruit).

The questions in this book are those I most frequently receive as Farm and Garden Editor of the *Miami Herald.* They come from gardeners as far north as Cocoa Beach on the east coast of Florida, and Bradenton on the west coast, and from throughout the Bahamas, Caribbean, and Central America.

Many of the trees and plants mentioned, however, have a much wider range, not only within Florida but throughout the worldwide tropics, in greenhouses, and in indoor situations.

No attempt is made to cover all of the 8,000-plus tropical plants that can be grown in South Florida under various conditions. But, fortunately, many plants share the same problems, symptoms, and growing requirements. If you are perplexed by a condition on a plant not named, check for a similar problem or procedure on other plants or trees within the appropriate section in this book.

Gardeners taking their questions to plant specialists can be more confident of an accurate answer if they give the most thorough description they can not only of a plant and its condition but of the care it has received. Samples of the plant are valuable — and not just one leaf, or one dead leaf, but several still-green leaves attached to a twig. These samples can be mailed to a county agent and will arrive in satisfactory condition even after several days in transit if they simply are put into a plastic bag and tucked into an envelope.

Acknowledgments

Thanks are due to the following for offering their advice, expertise, and time: Dade County Cooperative Extension Service agents, including Seymour Goldweber, who read several chapters of the book in manuscript and added his valuable observations to the chapter on fruits, his specialty; county agent Lou Daigle, lawns and ornamentals expert, who read the remainder of the manuscript; county agent Dr. William Stall, vegetables answer-man; Broward County agent Lewis Watson with his ready and detailed answers; lethal yellowing experts George Gwin, regional supervisor of the Bureau of Pest Eradication, Division of Plant Industry, Florida Department of Agriculture, and Dr. Randolph E. McCoy of the University of Florida's Fort Lauderdale research station; the South Florida Orchid Society; and Laurace Dickson, as usual.

The Florida Gardener's
ANSWER BOOK

Maintaining an attractive Florida garden is a pleasure for tropical plant enthusiasts. Response to any form of extra care is gratifyingly quick in most cases, and the fast growth rate makes do-it-yourself plant propagation an easy, inexpensive way to attain a lush look, often in just a year's time.

1. General Garden Care

For special index to this chapter, see back of book.

Gardening in Florida and the Caribbean involves a broad range of conditions that often vary from mile to mile, sometimes from block to block, and occasionally even from backyard to front.

Sand or rock, muck or marl, influence soil preparation and fertilizing; rainfall determines the amount of attention to watering, insects, and soil organisms; absence or prevalence of salt in the air (and in the water supply during droughts), temperature, exposure to wind and prevailing cold weather in winter, and elevation in storm weather affect plant selection, as do those significant variables of sun and shade.

Choosing plants and keeping them alive sounds baffling to the beginner not familiar with many of the thousands of exotics available, but there are handbooks that can remove the question marks. (See Bibliography.) Visiting nurseries and botanical gardens and analyzing other yards helps establish favorites whose suitability and care can be checked easily in one of these guides or through a County Cooperative Extension Service office, headquarters for free booklets and advice on all gardening problems in Florida.

Q. I need some advice on landscaping my home but I doubt that I can afford a professional landscape architect. Can you suggest any way I might begin to simplify this project?

A. The easiest way to start is at a nursery with one or more landscape designers on its staff.

A nursery designer does not command the price of a licensed landscape architect. (To use the term "landscape architect" in the state of Florida, the landscaper must be college-trained and have passed state examinations.) Nursery landscape designers are usually very flexible in the way they can work with clients.

For under $50 most can draw up a rough plan based on a sketch of your lot and its dimensions. It might indicate areas for trees, shrubs, ground covers, and grass, and it would suggest some of the plants you might like to use.

The more specific the plan is, of course, the more it will cost. Call up some nurseries and ask for the hourly consultation rate of their designers. Decide how long you can afford to listen to them and make an appointment.

The better prepared you are when you go into a consultation, the more useful the designer's suggestions will be. Are you interested in low or medium maintenance plants, a semiformal, jungle, or Japanese look? Are there unattractive views you want to block out? Where are the existing trees and plants you want to save? What are the traffic patterns across your yard?

You'll find books to help you round out the designer's plan in the Bibliography of this volume.

Q. I'm new here, just starting to plant my first tropical garden, and wonder what the life expectancy of most Florida shrubs is. One of my neighbors has had a croton hedge, she says, for 15 years. I know up north that many types of vines and hedges last much longer than this.

A. Generally speaking, the faster any life form grows, the shorter its life span, and this holds true for most tropical plants that are fast growers. (A notable exception: ficus.) Many exotic plants might need to be replaced after eight to 10 years. Past this period they tend to lose their attractiveness, get too big or too leggy, and, particularly if they are subject to repeated prunings, become gnarled or stunted in appearance.

Q. I've got a very sandy soil — the envy of my friends who are gardening almost on rock — but it still looks bad to me. What's the best way to make something out of it?

A. Ideally, if you were to take leaf mold, well-rotted manure, or peat moss, spread it four to six inches deep over your sand, and mix it in to a depth of nine inches, you would have an ideal planting medium for plants, trees, and lawns. The growth response of plants would be well worth the effort. Depending on your budget and your energy, however, you can scale down this ratio or prepare only special beds to this mixture. Don't mix in leaves, sawdust, or pine needles as conditioners. These materials need to be composted for some time before they are ready to serve as soil mixes. Without composting, however, they can be used as surface mulch.

Q. I have a never-ending argument here about the watering of lawns and of garden vegetables. Some claim it is okay to water on hot sunny days during the day. Others say water only at early morning, late afternoon, or on cloudy days. According to my understanding, lawns, plants, trees, etc., should have one inch of rain all at one time once a week.

A. In general, there's no reason why you should avoid sprinkling on a hot, sunny day. Some people hold to the theory that the sun's rays are magnified by the water drops and burn the plant. This doesn't take into account the lack of damage after a shower on a sunny day.

Getting water to the plants when they need it is the important thing. An inch of water per week applied at one time should take care of most lawn grass and established vegetable gardens. Of course, during excessively dry periods or in quick draining soil, more frequent watering may be necessary. Trees or shrubs with large root systems don't usually need water that often. A good soaking every two or three weeks may do.

Time of day for watering isn't terribly important. Sprinkling grass early in the day allows the blades to dry quickly. Watering late in the day would mean foliage stays wet for a longer period than is necessary. This could help fungus grow.

Q. What's the difference if I water my plants, say, once lightly every day or give them two deep soakings a week?

A. Conventional sprinkling by hand makes roots grow near the surface where they quickly dry out. The gardener who relies on a stroll around the yard with his hose had better keep walking because by the time he reaches one end, a good portion of his first application has already evaporated.

To see how much good your water does, push a trowel in the soil, and pull it toward you to open the earth. If it's moist to a depth of about four inches, you can roll up the hose.

Q. What's the difference between muck and marl, and which is the one that doesn't need fertilizer added to it?

A. Sorry, but some marl or muck enthusiast seems to have convinced you that rich color contains nutrients to match. Actually, according to the Cracker adage, South Florida soil generally is "just something to hold the plant up."

Marl soils retain moisture and are used almost exclusively for field nursery stock, root crops and other vegetables. It's a good type of soil for balling and burlapping plants, and with proper fertilizing and water control it will produce quality plants.

Muck, found in the lower glades areas, does not hold moisture and decomposes rapidly due to bacterial action. Although it looks like fertile soil, it contains only about two percent organic nitrogen. Many nurseries use a mixture of muck and sand as a planting medium. The medium is about 60 percent sand.

Q. I've read the explanation before, but will you tell me one more time what the term "pH" means? I know it has something to do with alkalinity.

A. A pH number refers to the degree of acidity (sourness) or alkalinity (sweetness) of your soil. So that the measure would be simple, an arbitrary numerical scale was adopted. At pH 7.0 a soil is neither acid nor alkaline, but neutral. Numbers below 7 indicate the degree of acidity; over, the degree of alkalinity. Native Florida soils range from pH 3.8 (very acid) to pH 8 or slightly higher.

The strongly acid condition is found in peats, muck, and palmetto flatwoods. Alkalinity is associated with South Florida marl and rock soils.

So what does the acid or alkaline content of your soil have to do with what's planted in it? It's a matter of chemical reaction.

Different degrees of alkalinity and acidity are required to "transmit," or make available to the plant, different nutrients. Sandy soil (on the acid side) can be changed to register a neutral or even slightly alkaline count by adding small amounts of lime to it.

South Florida's alkaline soils with their excessive calcium work chemically on manganese, copper, iron, and zinc so that plants can't get them out of the soil. So, these so-called trace elements, in alkaline soil, are best introduced to most plants through their foilage rather than by attempting to remodel the soil to make them acceptable. Exceptions are vegetables and palms whose roots are able to pick up trace elements by countering this chemical reaction.

Q. How can you tell an iron deficiency from a manganese deficiency? Or a boron deficiency from a calcium problem? Please describe the symptoms of the different deficiencies and what to do about them.

A. First, describing them makes treating them sound complicated. But, essentially, plants that show veined yellowing simply should be given a nutritional spray at least twice a year — at the beginning and near the end of the growing season, for example — and a couple of soil applications of chelated iron.

The presence of the leaf-greening elements in a bag of fertilizer does not make these separate foliar and soil applications unnecessary in South Florida soils because the formulations are often inadequate to cope with the high calcium carbonate content. In soils that do not have such a high pH, these additional applications of secondary elements are not as necessary.

Leaves lacking iron or manganese appear much alike. Both deficiencies cause new leaves to turn yellow, leaving the veins green.

When there is a lack of iron, only, the large veins remain green. When the plant is manganese-deficient, all the veins, including the smallest, form a network of green veins within the yellow leaf. Leaves showing a manganese deficiency have scattered dead tissue. No dead tissue is obvious on iron deficient leaves.

Zinc deficiency is called "little leaf disease" because terminal leaves get smaller, shorter, narrower, curled, or somewhat sickle-shaped. There is broad green veining on yellow as compared to fine veining with other deficiencies. The leaves usually thicken, and the veining on the underside of the leaf is more prominent.

Boron deficiency, a less common problem, shows itself in distorted and severely twisted new growth. The terminal bud and root tips die, and the new leaves yellow and break down at their base.

With calcium deficiencies, new leaves stay small and develop a hook. The terminal bud is severely twisted and soon dies. New roots turn brown and the tips are dead. Since the calcium level in South Florida's alkaline soils virtually is always adequate, potted plants are the most likely, although infrequent, victims of calcium problems.

Magnesium deficiency is sometimes called bronzing, indicating the appearance of older leaves on some trees with this problem. Often older leaves will turn yellow with a pointed "arrowhead" of green extending along the midvein. Leaves may also be colored reddish or orange.

Foliar sprays will clear up manganese and zinc deficiencies. They sometimes contain magnesium, too, but this element is more effectively applied as magnesium sulphate or Epsom salts in quantities ranging from a tablespoon for a small plant to a half pound and up to five pounds for a tree, depending on its size. Chelated iron is applied (with plenty of water) to moist soil.

Not all nutritional sprays contain boron, so be sure to check the label if that's your problem. Don't use boron purchased as supermarket borax, because this element is toxic to plants in any but very small doses.

Q. Just what nutrients does cattle manure contain? I can get all I want from a stable, but I have to haul it and need to know what else — if anything — will need to be added.

A. It takes 1,200 pounds of cow manure to equal the amount of nitrogen, phosphorus, and potash (the three numbers on a fertilizer bag) found in 100 pounds of a commercial 6-6-6 fertilizer. Or you could say that in 100 pounds of cow manure there is a half pound of nitrogen, one-third pound of phosphorus, and a half pound of potash. If you are interested in an organic approach and also in root crops, you can add steamed bone meal, which contains about two percent nitrogen, 27 percent phosphorus, and no potash, for a better balance.

Composted (be sure it's well rotted) manures are great nutritional soil conditioners, but, as you can see, if your soil is nutrient-deficient to begin with, you're going to be busy shoveling for some time. You can't go wrong with them, however; see how the manure works for you, and if it doesn't produce the kind of plants you want, it still will make a superlative base for the addition of fertilizer supplements.

Q. What's the best grade of fertilizer to use in Florida?
A. Most garden plants like a 1-1-1 ratio. A 6-6-6 fertilizer will do the job, plus a twice-yearly application of nutritional spray containing the minor elements that, due to South Florida's alkaline soils, must be introduced through the leaves. Chelated iron added to the soil helps with iron deficiencies here.

Q. I have been told that I could save money by using granular 10-10-10 instead of the soluble 10-10-10 fertilizer recommended when I bought my plants. I'd like to save money, but I'm still banking on my nurseryman, even though my neighbor has a green thumb.
A. Your neighbor omitted an important detail. The soluble mixture is the right one to use when the plants need a starter. It is quickly picked up by roots. The hard granules in a granular 10-10-10 take time to dissolve so are resistant to leaching and will hold through several showers, giving an advantage in the cost per pound when rain is frequent.

Q. How hard does it have to rain before you've lost all the fertilizer you applied a few days before a downpour? We had a real flood after I finished fertilizing our trees and shrubs, and my wife says I should do the whole job over.
A. In the future, if you're worried about leaching, put out a coffee can to measure the rainfall in your yard following an application. Generally, a three-quarter to one-inch rain will not wash fertilizer beyond the root zone. If it rains more than an inch, the amount leached beyond the roots will vary with the moisture content of the soil, amount of water, rate of water introduction, soil type, land shape, soil fertility, and other factors.

As a rule, on rock soil, a three to four inch rain over a one to two day period will leach all the dissolved fertilizer beyond the root zone.

Q. Can you give me directions for making a compost pile?

A. Choose an out-of-the-way corner of your lot. If you feel ambitious, you can construct a three-sided, U-shaped enclosure out of wood or block. Then just start piling grass, leaves, and other kinds of plant material layer on layer, turning it over once a week or so. In time, bacteria will go to work and break it down, leaving you a rich and crumbly collection of organics.

Always remove the material from the bottom. It is great for building the soil in beds and around individual trees and shrubs.

Fancier compost piles layer a foot of plant material under four inches of manure sprinkled with five pounds of 6-6-6 and watered well. The layering is repeated as new mulch is added; turning regularly is still necessary.

Q. Should you put fertilizer in the bottom of a hole when you plant a tree or shrub? What's the best way to prepare a hole?

A. Any fertilizer except Milorganite or manure is likely to burn tender roots if put in a planting hole, and it should not be mixed into the soil or placed in the bottom of the hole. Scatter it on the surface as usual.

There are several ways to prepare a good hole; however, all holes should be dug at least double the width of the container and six inches deeper than the pot.

Since fertilizers and water leach quickly through native rock and sand soils, peat moss or humus should be mixed into the material dug from the hole or, better, into nursery topsoil purchased in bags. Various combinations are beneficial: two parts peat moss to two parts topsoil and one part manure or Milorganite; one part topsoil to one part peat moss and one part humus; or two parts peat moss to one part native soil and one part manure or Milorganite.

The mixture should be poured into the bottom of the hole, watered, and the plant set in so that its crown (where the trunk meets the soil) is no lower or higher than it was in the container. As the soil is poured in around the roots, it should be watered with the hose turned on to just a trickle. This will help settle the soil and eliminate air pockets. It will also make prolonged watering unnecessary when you are finished.

Milorganite or manure may be used on top of the soil, or a light application of chemical fertilizer. An inch or two of mulch is helpful around the tree to help conserve water. Water daily, if possible, for the first couple of weeks, then twice weekly to help a young plant get established.

Q. I'd like to make my own root feeder since I've read that getting the fertilizer right down in there helps roots pick it up faster. Any suggestions on how I might go about this?

A. There are a couple of simple approaches, but be sure you use a liquid fertilizer, supplemented by feedings with chelated iron and a nutritional spray a couple times a year.

With one version, you buy some PVC pipe at the hardware store, drill holes around it — maybe a dozen quarter-inchers — and dig a hole for the pipe, inserting it permanently. It can be anywhere from eight inches to a foot deep. Then you fill the pipe with liquid fertilizer at each feeding.

Or you can buy a lawn drill aerator. The Alsto Company (11052 Pearl Road, Cleveland, Ohio 44136) sells one version. Two water jets penetrate the soil as its two stainless steel spikes attached to a waist-high handle are pushed into the ground. You can connect this to a hose — or hose-end sprayer which has been connected, of course, to a hose.

Root feeding eliminates water runoff, too. There are some spray companies who specialize in root feeding with special formulas that contain the nutrients you otherwise would have to apply separately, such as chelated iron and nutritional sprays.

Q. I read in a garden book recently that you should toss your tea leaves, coffee grounds, and egg shells around your plants to give them extra vigor. I thought that this was an old wives' tale. Is there really any value in using these discards as fertilizers?
A. These materials will add humus and probably help — slightly — to improve the "mechanical" condition of the soil. They aerate the soil, and the tea leaves and coffee grounds will help conserve moisture, as will any sort of organic mulch such as grass clippings and leaves.

The egg shells are a source of calcium, an element that plants need to make strong stems and good normal growth. The only drawback is in relying on these kitchen castoffs as plant food. Few people can apply tea leaves, grounds, or egg shells in quantities large enough at a time to do much good.

Q. I've brought in a load of mulch to give my plants a nice warm blanket for the winter, and now I hear that mulch is harmful to plants. What's the story?
A. Depends on where you put the mulch. If, for instance, you put it over a tender ground cover such as wedelia just before an expected cold snap, it will hold warm air, rising from the ground, around the plants. On the other hand, if it covered only the base of a plant, it would prevent heat waves from rising and increase chances of injury from cold. A mulch used in this way will keep the plant more vigorous, up until a frost arrives, by conserving soil moisture, but it definitely should be raked back from under any part of the plant the night a frost is expected.

Q. I have a neighbor whose washing machine is now running water over into my garden. How upset should I be? I don't have my prize plants on that side but I don't want my hedge damaged, either.

A. Most sudsy water is quite alkaline. Dade's naturally alkaline soils do not need more alkalinity. Soapy water is damaging to good soil structure. Constant flooding with the bubbly mixture will make it less porous and will cut off circulation of air in the ground.

Q. I have a three-and-one-half-gallon compressor sprayer and have a lot of trouble getting my sprays on the leaves evenly. I use a sticker-spreader, but the sprays seem to drip and puddle up. I get some results but very little.
A. It sounds as if you may be applying the spray too heavily. The idea is to apply spray only to the point where it runs off the leaf. In general, it is a good idea to try for good coverage on the undersides of the leaves. Enough spray will fall back onto the tops for effective action.

Q. When is the best time to transplant trees and shrubs in Florida?
A. One guideline says deciduous woody plants in December, broad-leaved evergreens in March or April, and palms in June. Even the experts differ, though, and different climatic zones in Florida call for slightly different dates.

Gardeners have had success at any time with all types, of course, but if the plant you want to transplant is on the fragile side, spring is safest. Transplants of tender greenery should be made in time for them to establish themselves well before a cold snap. Also keep in mind that transplanting may adversely affect flowering or fruiting if not done far in advance of the blooming period. Successful transplanting is very dependent on the quality of the transplanting hole and devotion to regular watering for the first month or so afterward.

Q. Without calling in a professional stump remover, how can I get rid of an unattractive stub in my front yard?
A. You can buy preparations that will gradually destroy the stump. Burning it out and pouring gasoline into cuts in the stump are other ways. You select the side with the fewest main roots and dig a hole along the side of the stump. High water pressure from a hose can help undermine the roots; the hole should be several inches deep. If the stump does not have a tap root, you can use the "long squirt" setting of the nozzle on the hose to drill a draft hole underneath the stump to the opposite side of the fire hole.

Next, start a small, hot, smokeless, smoldering fire in the bottom of the stump. To speed it up, you may want to crunch fire coals down against the stump roots. Never leave it unattended. When you want to go away, put it out and start over at a later date.

Stumps resist all attempts at beautification unless they are located in a position appropriate for a flower bed or hedge to engulf them. They don't make the grade as planters, either — the effect is either artificial-looking, out-of-proportion, or slightly grotesque, though you might consider trying to arrange a cluster of areca palms around the stump.

Q. Would you give me some guidelines on preventing tree damage to my house in case of a hurricane? I'd like to avoid drastic trimming if possible.

A. Actually, a 100-mile-per-hour wind is not as dangerous as it sounds if necessary measures are taken before the wind reaches gale force. Here is a check list to help avoid damage from trees:

1. Remove all dead branches and masses of moss that could be whipped off the tree in a soggy mass. Take off branches that could batter your house.

2. Check for rotten spots. Fan-like fungus growing from the area is an indicator. So is damage by woodpeckers, earlier storms, or a car bumper. Clean out and fill with a watertight material.

3. Some older trees with Y-shaped crotches are inclined to cleave apart. Prop temporarily or make the tree a unit again with steel rods and turnbuckles.

4. Remove loose palm leaves and coconuts.

Q. What can I do for ixora and sago palms that were nipped by frost during the past cold spell?

A. Be sure that cold-damaged plants do not lack water. Damaged foliage does not wilt as normal plants do when they are too dry. So, irrigate them well each week during the recovery period. Don't be in a hurry to remove damaged foliage. The damaged parts can give some protection for the plant in case of another chilly spell.

Q. We've just finished building a vacation home in the Keys and now we're ready to plant. Some of our neighbors have avoided doing much landscaping around their houses because they say salt is too much of a problem. We're determined though to make our cottage an "island paradise" without too much work — if it can be done. What do we have to choose from?

A. Here are a few of the trees that give good results on the Keys and in other coastal spots: coconuts, acacias, seaside mahoe *(Thespesia populnea)*, called "cork" in the Keys; tree hibiscus *(Hibiscus tiliaceus)*, Jerusalem thorn *(Parkinsonia aculeata)*, sea grape, black olive *(Bucida buceras)*, silver buttonwood *(Conocarpus erecta)*, gumbo limbo *(Bursera simaruba)* and, of course, the controversial Australian pine or *Casuarina equisetifolia.*

The casuarina will grow without any care at all on the poorest beach sand and with its feet in salt water. It's a good soil builder and windbreak. For home use, it becomes objectionable only when it attains considerable size. When the ground becomes soggy and the wind strong, it may topple because of its shallow root system.

Three acacias are native to Florida. *Acacia cyanophyla* flowers abundantly on the beach above Fort Myers at Boca Grande. *A. longifolia* and *A. retinodes* are cold-hardy trees for more northerly coasts.

In addition to the Australian pine and the acacias, the three trees listed that will bed down right on the waterfront are the sea grape, silver buttonwood, and seaside mahoe. The others should be planted farther back.

Among the shrubs and smaller plants that like Keys living are Spanish bayonet, century plant, cocoplum, oleander, and lantana. The Spanish bayonet and century plant make good screens to protect less tolerant plants.

You can even have flowers in salt-sprayed areas. Periwinkles, gerberas (African daisies), gaillardias, wedelia, the beach sunflower, and portulaca are all highly salt tolerant. There are lilies to choose from, too — such as the American crinum or swamp lily, lily turf, and daylilies.

Q. I'd like to know, once and for all, whether South Florida is tropical or semitropical.

A. Pioneer Florida horticulturist-forester Dr. John C. Gifford classified southern Florida's climate as "tropical-wet-dry," identical with patches of the West Indies, Central and South America — a microtropics on the American mainland, distinct from any other climatological region in this country.

Another prime indicator of the tropics used by Gifford and cited in *On Preserving Tropical Florida,* a University of Miami Press anthology of his writings, is the presence of the coconut palm and other native and introduced tropical species.

Among his contemporaries who also claimed South Florida as tropical was David Fairchild. He stood behind the coconut palm indicator, also. Thomas Barbour said the presence of tropical spiders cinched it, and Charles Torrey Simpson defined as truly tropical any area where "a majority of the native plants have been derived from the Torrid Zone."

Ornithologist A. H. Howell cited the many tropical birds common to southern Florida as proof.

Q. I realize it's too late to help me now, but I'd like to know what happened when I tried to remove house paint from my monstera. The painters missed it with their drop cloths, and it was very spattered with paint. After I removed the paint with solvent, the leaf under the splatters turned from light green to brown and the brown gradually spread. Is there anything else I could have done?

A. Different paints have different effects on plants — all of them bad and none of them reparable. Oil-base paint does more damage than vinyl water-base paints, which usually leave the area around the spots unharmed. Spray paint generally lands on leaves in a drier state and may or may not be worn off by the elements.

Solvents used to remove paint will only penetrate into the living cells and increase the damage. Attempting to chip it off by hand often results in exposing injured tissue to sunburn and removing cells, which results in the death of adjacent tissue.

Q. What's the best, all-around spray for control of bugs in South Florida?

A. No one spray will get rid of all pests, and it may often be necessary to make additional treatments with other insecticides. But an example of a good general-purpose spray is one that contains (1) Meta-Systox-R, Malathion or Diazinon, plus (2) Sevin, plus (3) Kelthane or Tedion.

One of the first three ingredients mentioned will discourage sucking insects, including aphids, mealybugs, whiteflies, and scales. Sevin controls a wide range of aggressive chewing insects like beetles and caterpillars, and Kelthane or Tedion goes after spider mites.

Labels should be carefully read and thoroughly understood. Call your county agent if there is any question. And, remember, chemicals can't always be mixed for convenience; your agent will know those that can.

Q. What do you use to get rid of those munchers that roll themselves up in leaves and then go to work on your plants? At first I thought they were only going to spin cocoons, then I finally realized they simply didn't want me to see the damage they were doing!

A. Leaf tiers and leaf rollers are caterpillars that roll and tie foliage together with strands of silk not only for protection against you but from birds. Go after them with Sevin.

Q. My garden is being attacked by aphids now, but there also seem to be quite a few ladybirds around. I've been told that if I spray with Malathion, I'll kill the ladybirds. Is there any alternative?

A. The beneficial ladybird beetle is a valuable aid in aphid, scale, and mite extermination and should be encouraged to multiply in every garden. Use oil emulsion instead of Malathion, and you will have two approaches to control working for you.

Q. There are hordes of these black grasshoppers in our yard with yellow stripes down their backs eating the edges of everything we've got. How can we get rid of them?

The lubber grasshopper is a colorful, voracious garden pest feasting mainly on members of the lily family from summer through fall. The mature lubber is red, orange, yellow, and black-spotted. The immature insect appears in swarms in the spring. It is black with a yellow or red stripe down its back. Trapping and discarding in coffee cans is best control.

A. Those are the immature lubber grasshoppers that come up out of the ground every spring.

You can try spraying Sevin, a stomach poison, on the leaves of their favorite plants — members of the lily family such as spider lilies, crinums, and amaryllis. Most people go after them with the heel of their shoes.

If this isn't your style, there's another way. When quite small, these pests are very nervous but when they reach an inch or so in size it is quite easy to take a coffee can and knock them into it with its lid. Save several cans for this purpose, and when you get a dozen or so in each can, take them for a ride and dump them in an undeveloped lot somewhere.

Q. The roof on my house has a large overhang, and I've noticed that occasionally some of the plants under it (mostly ferns) show damage after I spray my roses nearby. There's no way I can cure the roses without killing the ferns. Or is there?

A. Some pesticides, particularly Malathion, may cause injury to certain plants if they are applied during the hottest part of the day or when sufficient plant moisture is missing.

Malathion can even burn some varieties of roses, including 'Caledonia.' And it has been known to damage Boston, maidenhair, and pteris ferns. It would be a good idea to give your ferns a good watering one or two days before applying the pesticide since your roof cuts them off from normal rainfall.

Q. What are the flat, oval, see-through "things" attached to the flip side of my sick leaves?

A. The larvae, or young, of the whitefly. Some people look for controls for the pest when it's a small, snowwhite, gnatlike adult, but the best time to spray for them is three weeks after the adults are seen. By this time the young have hatched.

Whiteflies flock to the strong, new flushes of growth produced by midsummer weather. They are well known to owners of fruit trees, gardenias, and other plants.

Before you reach for a spray, give Mother Nature a chance to get rid of them by permitting the "friendly fungus" (red and yellow aschersonia are two of these entomogenous fungi) to attack the nymphs as they develop on the leaves. These fungi may be "large," up to a quarter-inch in diameter, and often are mistaken for pests. They can be distinguished from scale by their two-toned color, whereas scale insects are usually uniform in tone.

Several good materials that will give effective control include Meta-Systox-R, dimethoate (Cygon or Defend) and Malathion, used according to container directions.

Q. I was told that my lantana is infested with spider mites and that the way to get rid of them was by spraying with Aramite. It may have set them back slightly but I can't see any real improvement. What now?

A. Some people find sprays disappointing because they don't know how to apply them correctly. Thorough coverage is very important, and in many cases that means the underside of the leaves, too. Those container directions are as crucial for success as a cake recipe is for achieving the desired results. (More so, if you consider that many pesticides, casually used, are toxic to humans.)

Q. Are these bugs harmful that have two glowing eyes and fly around at night? I'd also like to know what they are.

A. Those "eyes" are really fluorescent spots above the true eyes on the head of the click beetle. The beetle takes its name from the sound it makes when it is placed on its back, then leaps into the air and turns over.

The beetle is harmless to both man and plants in this adult stage, but in the immature larval stage it is known as the wireworm, a damaging pest on flowers and vegetables, particularly corn. It usually attacks either seeds or very young plants.

Q. I am exceedingly annoyed by the presence of snails in my garden. They crawl over the outside walls of my house and probably devour many of my plants. They appeared about three years ago after delivery of some top soil. What can I do about the situation?

A. These little critters are indeed a nuisance, but they can be controlled and eliminated. Use any of the commercial baits from garden shops, or beer cans tilted on their sides. The residue of beer acts as a lethal dessicant.

Make sure, first, that these snails actually are eating your plants (they will leave faint silver trails, usually). Not all species of snails are plant pests.

Q. There are some spiders in my garden. I'm afraid to work around them although I've been able to examine them closely. They are covered with reddish-bronze hair and have bluish mouths. What can I do to get rid of them?

A. Sounds like it might be phidippus, the jumping spider. No need to worry — phidippus bites flies, not people. She is harmless and beneficial.

Q. Will you please advise the correct method to be applied in destroying toadstools and mushrooms?

A. You could drench the area in which you find the toadstools and mushrooms with a fungicide such as neutral copper or one of the general purpose lawn fungicides. Mix at the rate recommended on the label.

You may have to repeat this several times as new growths appear. The growths often are confined to a relatively small area, and they may rise from a piece of wood buried beneath the soil or lawn surface.

Q. I know frogs and lizards have their place and that they were here first. But in the rainy season the frogs keep my wife awake at night, and I don't like the idea of lizards occasionally slipping into our baby's room. Is there any safe way to discourage them?

A. If it's any comfort, Florida does not have a poisonous lizard, and the harmless ones prefer window sills to humans. Noisy frogs are best coped with psychologically by accepting their "song" as you would rain on the roof, crickets, or any other normal environmental sounds. (Some of the same people that say frogs keep them up can sleep quite easily with the roar of an air conditioner.)

Getting used to these temporary "guests" is safer than trying to bait your house with poisons. If you must do something, you might spray the shrubbery around your home for insects, thereby reducing the lizard and frog population by decreasing their food supply. Frogs should be regarded as assets because they eat a great number of noxious bugs.

2. Lawns

For special index to this chapter, see back of book.

Do you want it to look like Astroturf or will you settle for less? Sometimes the decision is taken out of your hands when a lawn problem is allowed to run its own course. Sprawling weeds, ravenous bugs, and subtle fungus growth can make a lawn an untidy blot on a neighborhood in relatively short order. Replacing large portions of a lawn when you could be buying outstanding trees and shrubs for the same money can be avoided.

Selecting and maintaining the different types of lawn grasses often perplex homeowners, but there's plenty of expert advice available on the subject. Check with your county agent for free literature on lawns, and establish good relations with a reliable nursery or garden shop specializing in lawn care products.

Q. I am trying to decide on the relative merits of grass versus gravel or ground cover. I'd like to have less grass to worry about, but I know you have to watch out for weeds in gravel and that ground cover has to be clipped and cared for. What are some of the advantages and disadvantages?

A. A combination of all three or at least two is not only more interesting to view but has maintenance advantages. You don't have to do something to all three at the same time, whereas if your property were covered in grass, the entire area would have to be mowed or fertilized at one time.

Gravel can be used in those shaded areas where grass grows poorly. Weeds won't be much of a consideration in a shaded area covered generously in gravel.

Lawns can be fancy or plain, but there's no doubt that grass regularly fed and watered is far more resistant to weed and bug invasions. Saint Augustine, bahia, and centipede grasses are most popular.

Grass is fine for open areas in full sun that are convenient to keep regularly watered. In theory, reducing the size of a lawn should make it easier and cheaper for you to keep it in peak condition.

Ground covers sold in nurseries need little care and are rarely fertilized after a good bed has been prepared with topsoil and, perhaps, manure. They do need to be kept in bounds. For some, a monthly pruning in the winter and shaping up every other week in the summer is desirable.

Where gravel meets ground cover there is little need for a mowing strip or divider since pruning shears will keep them apart. But between ground cover or gravel and grass you might want to lay bricks, small creosote posts, or other devices to keep them apart.

Q. I'm new to South Florida and baffled by the different types of grasses. What's the best kind for a lawn here?

A. St. Augustine is the most popular; it's estimated that at least 80 percent of the lawns in South Florida are planted in this grass because good sod is available and it grows vigorously. Its major drawback — chinch bug susceptibility, although more chinch bug-resistant strains are becoming available. There also is a new variety, 'Floratam,' that is resistant to the St. Augustine Decline virus not yet in Florida but damaging lawns elsewhere.

Chinch bugs stay away from Bermuda grass, but it's a high maintenance blade that needs more fertilizer and more frequent mowing. Zoysia is that very fine textured, wiry grass. Its tough blade requires a reel type mower. Zoysia is liable to various diseases and nematodes, but many people like its neat look and grow it successfully.

The Bahia grasses do best on the Florida west coast and in central Florida although some grow it here, too. It's a low maintenance grass but makes a "thinner" lawn. It needs a rotary mower.

Q. I am building a new home and am wondering about using Argentine or Pensacola Bahia for the lawn. I've always had St. Augustine and like the appearance, but I'm tired of fighting the bugs. Is there a low maintenance grass? How about the practicality of these grasses?

A. The Bahias are low maintenance. The coverage is not as adequate as St. Augustine, but they don't attract bugs. Yet they present a mowing problem because they put up long seed spikes requiring a rotary mower. In the summer they require frequent mowing.

Centipede grass, while requiring plenty of moisture, grows well in South Florida on acid soils. It has a tendency to turn brown during the winter. But it requires less mowing. When setting a new lawn, it is best to consult a reliable nurseryman or turfgrass expert at a County Cooperative Extension Service office. Your specific needs in relationship to the soil you have are better met this way.

Q. Would you tell me if Pensacola Bahia grass will gradually fill in bare spots in my lawn. My lawn was seeded to Bahia. It came up spotty so I loosened soil in bare spots and put in seed at least two times, but nothing happened.

A. Since Bahia does not produce runners like St. Augustine, it does not spread rapidly. I wouldn't count on it to cover any sizable bare spots. Another try at seeding probably is your best bet, following the routine of keeping the seeded areas moist until the seed sprouts and the new plants become established.

Q. I have a scraggly-looking lawn I'd like to fill in with some type of fast-growing grass. What's the best and how long will it last?

A. Several different grasses could be used, including bentgrass, bluegrass, and fescuegrass, but rye is the most generally used and easiest to obtain.

Overseeding in November will give a good temporary lawn while the permanent grass is recovering from the ailments of the summer. It should last until March or April.

To prepare the lawn for overseeding, the grass should be mowed very closely, the excess clippings raked up, and the soil loosened a little so the seed will come in contact with the soil. Lightly fertilize with a 6-6-6 or 8-8-8 fertilizer and water it in before seeding.

Water should be applied lightly to the seeded lawn once or twice a day until the seedlings are well established. Subsequently, the grass can be watered as needed and fertilized lightly but frequently to maintain it in healthy condition.

Rates for seeding are: ryegrass, five to 15 pounds per 1000 square feet; bentgrass, one to two pounds per 1000; bluegrass, two to four pounds; fescuegrass, three to six pounds. The low rate should be used where minimum coverage is desired and the high rate for maximum density.

A combination of grasses is also practical, in which case the amount of seed should be adjusted accordingly. (If two kinds of seed are used, divide each of the recommended rates by two.)

Q. After you fertilize your lawn, does it really matter how soon you mow it? A neighbor of ours who lives for his lawn says his nurseryman recommends waiting a week.

A. A study by University of Florida researchers has shown that it takes six days for an application of fertilizer to result in the greatest benefit to your lawn. It has to travel up into the blade to be converted into the starch that feeds the rest of the plant. But if the blade is whacked off prematurely, what you've done in effect is served the rolls and salad but tossed the main course on the trash pile.

Q. My son and I like to do our own yard work but we don't know too much about the care of grass such as spraying and fertilizing. What methods are best for the layman?

A. It is not possible to give you a complete guide for lawn care, but a lawn should be fertilized three or four times a year. Use a 6-6-6 or 8-8-8. High-nitrogen (the first number) lawn fertilizers should be used only on large lawns, away from fruiting and flowering trees. It can adversely affect the production of flowers or fruit.

If you can find a 2-1-2-1 or similar analysis, this would be an excellent choice for sandy soils.

Your garden supply store can advise you on the sprays that are available for chinch bug, brown spot fungus, etc. You might also contact your county agent's office and request a booklet on lawn care.

Q. We are concerned with the proper maintenance of St. Augustine grass. Some folks say it is better to leave all clippings on the lawn to decompose and form a mulch. Others say it should be raked and removed to prevent formation of a thatch that becomes a haven for bugs.

A. Fertilizing St. Augustine grass should be done three to four times a year using a 6-6-6 mix. Watering should be once or twice a week except, of course, during abnormal rains. It is better to rake or bag all clippings. Left on the lawn, they do help increase the buildup of thatch, offer a good hiding place for destructive bugs, and make control of fungus more difficult.

Grass clippings should be placed on a compost pile where they can decompose and later be used to build up the organic content of flower and shrub beds.

Q. When should I apply lawn fertilizer in the fall? Should the application be heavy or light? If heavy, would one application suffice for the entire year?

A. An established lawn should be fertilized four times during a year: February or March, April or May, early September, and November or December. In the fall, use a complete fertilizer of 6-6-6 or 8-8-8. Use 12 to 16 pounds for each 1,000 square feet. It would not be advisable to use one heavy application in the fall.

Q. Have I been doing the right thing to pull out the unsightly runners in my lawn? A sprayman recently told me I was making a mistake.

A. Although you don't say, you may be referring to St. Augustine runners. Very often this grass will put out runners that just seem to grow loose on the surface of the lawn. If this is the case, they should be clipped. If the runners are moving into flower beds, they should be edged. If the lawn has a bare spot or a sparse growing area, however, the runners should be encouraged.

Q. We use a rotary mower on our lawn, and for eight years have never had any trouble. We mow every week. The grass is St. Augustine.

For the last four weeks, every time I mow it, it is scalped. I am watering the lawn twice weekly. Can you suggest something to prevent scalping? Would a load of sand help?

A. It is quite possible that thatch (a layer of old runners and clippings) has built up on the soil. Now the weight of the mower makes the wheels sink in to the turf and the blades scalp the grass.

Sand is an answer, but not a very good one, particularly in October. Applied then, sand often seems to promote fungus growth in the grass. The best time to sand a lawn is spring, say April or May, when the season of fastest growth is beginning.

It is possible to have thatch removed by firms which have special equipment. The process is called verticutting and is done with mowers that remove the spongy material.

Over a period of years, frequent sanding builds up the level of the lawn above walks and other structures. Verticutting is done by very few firms and the cost is not negligible.

As a temporary measure, if you have not already done this, be sure your mower is set for its highest possible cut.

Q. What can I do about the grass that's yellowing over my septic tank? I understand it should be sprayed with some special material.

A. Apply iron sulphate or neutral iron to the area, according to container directions. Be careful not to get it on sidewalks or walls — it stains.

Q. I'm taking a good deal of kidding from my friends because I'm still using the reel mower I bought from the former owner of our house. He said this was the best type to use on zoysia and since I was a newcomer I took his word for it. Maybe you can give me a better answer for my friends' quips.

A. Both zoysia and Bermuda grass should be cut with reel-type mowers, which have the cutting action of scissors. This cutting action prevents damage to the grass. The important point is to use a good mower with sharp blades. Any type of lawn grass can be severely damaged by dull blades that tear or beat the grass instead of cutting it cleanly.

Q. I get yellowish-greenish patches in my St. Augustine that turn to brown, and I suspect the problem is chinch bugs. This is my first St.

Augustine lawn, and although I've rooted around in it, I can't see any bugs. How can I test for them and what's their control?

A. Take a coffee can with both ends removed, push it a couple of inches into the ground in a suspect area and fill it with water. Wait about five minutes and, if they're present, you'll see the little bugs floating on the surface. Then take your pick of controls: Diazinon, Trithion, Ethion, Zytron, or Aspon. A second treatment should be applied about 10 days after the first since eggs present often escape the first spray.

Grass cut too short too soon after fertilizing will rob the lawn of nutrients made by the blades. Grass allowed to grow too long will be sunburned after it's mowed and tender lower blades are exposed.

Q. I think my St. Augustine grass has something the matter with it that it isn't supposed to have — dollar spot. I know this is a problem with Bahia and zoysia grasses but my St. Augustine has the classical symptoms — soft brown patches that eventually die completely. What could it be?

A. Dollar spot isn't as exclusive an ailment as you believe. During periods of warm days and cool nights it will attack Bermuda, St. Augustine, centipede, and Italian ryegrass, as well as Bahia and zoysia. With the last two grass types it identifies itself first as small, two-inch circular areas bleached to a straw color.

On Bahia grass it attacks fairly large areas. Spray with Daconil, or Fore, covering the entire lawn to keep healthy areas from being infected. Repeated sprayings are usually necessary. Applications of nitrogen will retard the disease.

Q. Our neighbor is having trouble with his grass. From the sample, can you give us some hints at getting rid of the problem?

A. Your neighbor is being troubled by gray leaf spot. These are specific fungicides for this problem. Ask your garden supply store for a wide spectrum spray. There are several that will give excellent results.

Q. I've got strange-looking crickets about an inch-and-a-half in length in my lawn. What are they and how can I keep them from doing more damage?

A. Mole crickets are most destructive in newly seeded and sprigged lawns but can be a serious problem in established grass as well. They also will damage tender ornamentals, young vegetables, and seed beds.

Safest control is with Kepone wettable powder or Diazinon emulsifiable concentrate used as a spray according to directions. Kepone, Dursban, or Dylox also come in baits.

Q. In the past few weeks we've noticed a small cloud of moths in our grass. What are they up to and should we do something to them?

A. Sod webworms, spotted greenish lawn pests, start their careers as small, dingy brown moths with a wingspread of about three-quarters of an inch. The moths lay eggs that hatch in about a week and start nibbling on your lawn. They're a problem for the following three weeks or so, feeding at night and lying in a curled position on or near the soil surface during the day.

They are most destructive when, as caterpillars, they reach the three-fourths of an inch size. Injured grass has notches chewed into the sides of the blades, which are eaten back unevenly. The grass may be almost completely stripped off in patches, and these close-cropped areas have a yellowish to brownish appearance. The caterpillars' webs show up easily early in the morning when the grass is covered with dew.

A week after sod webworm damage is first spotted, spray with Sevin (50 percent) wettable powder at the rate of 1½ cups (in a quart-size hose sprayer) per 1,000 square feet of lawn, or Diazinon 2E emulsifiable concentrate, three-quarters cup per 1,000 square feet of lawn. Repeat in about two weeks.

Lawn sprays commonly sold by garden supply stores usually contain a mixture of two or more insecticides effective for controlling these chewing caterpillars, as well as chinch bugs and other pests.

Q. I understand that earwigs, those bugs that look a little like scorpions, are harmless. True?

A. To people, yes. To grass, no, since occasionally they will feed on tender roots. They are difficult to control, and repeated applications at two-week intervals may be needed. Use Sevin at the rate for sod webworms and Diazinon or Baygon at the rate for chinch bugs. Amounts and instructions will be on the container.

The mixture may be made more lethal by adding 2 tablespoons of BHC wettable powder or one and one half teaspoons of lindane emulsion concentrate (15 to 20 percent) along with five tablespoons of Sevin 50 percent wettable powder per gallon of water.

Q. I've got a lot of weeds I'd like to get rid of. What's a good weed killer?

A. Depends on what you wouldn't mind getting rid of in addition to your weeds. There are some that kill any plant they touch. They are valuable for areas such as gravel or shell driveways. Cacodylic acid, a chemical that doesn't persist in the soil, performs this job well.

A 2, 4-D product is effective on weeds and does not seriously injure most grasses. Atrazine is frequently used on St. Augustine, centipede, and zoysia (but not Bermuda, Bahia, and carpet) grasses.

Q. A grassy weed is taking over my yard and it seems I can dig shoulder-deep and still can't get all the roots. Is there help for this problem?

A. This sounds like torpedo grass. If it is scattered throughout the lawn rather than being located in one area, you are faced with a tough problem. If you are willing to sacrifice the lawn, use Kilz-All. It is nonselective and will kill everything. The application should be repeated within two weeks. If the lawn is involved, sodding will give the best results.

Q. My beautiful 'Bitter Blue' (St. Augustine grass) lawn in my front yard is being completely ruined by nutgrass, which practically has taken over. Is there anything I can do to kill this grass without affecting the good grass?

A. Any weed killer potent enough to knock down the nutgrass would cause severe injury or kill the St. Augustine grass. This lawn grass is particularly sensitive to weed killers.

Keeping the 'Bitter Blue' in good condition will help crowd out the weeds, and regular mowing will keep the nutgrass from being terribly apparent.

Q. What will kill so-called water sedge that is spoiling my lawn?

A. The "sedge" is a member of the nutgrass family. If it is in St. Augustine grass, I suggest you fertilize with a mix containing a herbicide, but not if you have fertilized recently. Two or three applications of the mix should greatly reduce the infestation.

Q. Will you please tell me how to get rid of Bermuda grass growing in my St. Augustine lawn?

A. Aside from pulling it up there aren't many alternatives. Often St. Augustine will thrive to a point where the Bermuda is "choked out." If it is widespread, hold back fertilizer and water. Bermuda is a "high maintenance" grass and it will be weakened with less nutrients, thus permitting the St. Augustine to spread faster.

Q. I am having a hard time getting rid of sandspurs in my yard. Is there any way to eliminate them?

A. Sandspurs grow under poor soil conditions. They do not compete well with lawn grasses, so if you can keep your lawn in a vigorous condition, this will help.

There are two ways of attacking the problem directly. Dig the pests out by hand or use a non-selective week killer that will kill all vegetation where it is applied. The effect of the weed killer will wear off after several weeks and you can plant sprigs or plugs of new sod.

Q. Most people want to know how to grow grass, but I want to know how to kill it. I cleared a strip in my yard, covered it with 90-pound asphalt roofing paper, punched holes in it for drainage, and then covered the area with three-eighths-inch crushed rock. It looked beautiful — until the grass found every hole in the asphalt paper. What's the answer?

A. You can periodically treat the area with something lethal like the product called Kilo All oi one that contains 2, 4-D, but since one of your objectives in covering the area with gravel probably was to make it work-free, better start over. Replace the asphalt with the plastic sheeting commercial strawberry growers use (available in farm and garden supply stores), and re-cover with the gravel. If you find you absolutely need holes for drainage, make them little more than pin pricks.

Q. Is there any way to kill zoysia grass roots so they will not interfere after we replant with centipede? We are digging the zoysia out by hand.
A. If you are planning to resprig the entire lawn, your problem isn't too severe. Careful work with a hoe and rake (not the fan type used for raking leaves) should bring out a lot of the roots. Then treat the soil with a combination fertilizer-weed killer before planting the centipede.

Q. We left our once-beautiful lawn in the care of a yardman for over a year and came back to a brown, burnt-up 6,500 square feet of sod with an immense covering of matchweed. On this we sprayed "Weed Be-gone" with good results. We were left with a brown, burnt-up, 6,500 square feet of sod without an immense covering of anything.

We then covered the entire area with a two-inch layer of silica sand and some pieces of St. Augustine, closely spaced. The St. Augustine took, establishing itself with a lot of healthy runners, but apparently St. Augustine doesn't travel alone. It also brought along a full retinue of assorted weeds.

On the advice of a local nurseryman, we fertilized with his lawn greener and builder. We find no fault with his advice or with his product. He felt the organic enrichener would choke off the weeds. It didn't; it mildly subdued them over a three-month period.
A. Back to the beginning. Although it isn't of much help to you now, the surest way to avoid a long drawn-out hassle with weeds in the future is to start with high quality (preferably, certified) weed-free sod or seed and properly establish it on soil that is not infested with perennial weed seed or has been treated with a weed killer well in advance of grass planting.

Grassy weeds in St. Augustine, centipede, and Bahia cannot be controlled safely with selective weed killers. It is better to use spot applications of Dalapon or Thytar on unwanted grassy weeds, being careful not to treat areas of good grass.

The only practical way to treat large areas of grass infiltrated with broadleaf weeds is to use one of the "weed and feed" products that live up to their names — but not overnight.

Weeds are more susceptible to weed killers when they are young and growing actively, so don't expect to see the big ones disappear other than by hand-pulling.

Q. We sanded our St. Augustine not too long ago because it was just getting too thick. Now we see some big brown spots on the lawn. Could this be caused by something harmful in the sand?

A. Many people will continue to sand their lawns despite advice against it from the County Extension Service. Best moves to make when St. Augustine gets too thick are either to tear it out and resod or to hire a verticutter to renovate it at about three-quarters of the cost of resodding.

Sanding encourages the development of a fungus disease called "brown patch." This problem causes the grass to be partially killed in more or less circular patches that begin as small spots and may expand to several feet in diameter. It may also cause a thinning of the lawn over a large area, and root rot may be obvious. The fungus which causes this disease is soilborne rather than carried through the air.

Brown patch may be controlled with Daconil 2787, Fore, Tersan 1991, or thiram. Use according to directions on the container. The material should be sprayed on the area so that there is complete and even coverage. Use enough water to wet the grass thoroughly. Do not mow or water the grass immediately after applying a fungicide.

Brown patch may not become severe if the weather is warm, sunny, and windy. Many times the fungus condition will clear up during sunny weather conditions without benefit of fungicides.

Q. Two questions on chinch bugs: Can you apply the killer in dry weather without hurting the grass, and can I spot-treat for them or do I have to do the whole lawn?

A. It isn't necessary to have a watered lawn before granular insecticides are applied, but if you spray, you should run your sprinklers for about an hour beforehand. This is to aid in the spread and penetration into thick mats of turf.

It's best to treat the entire lawn thoroughly, concentrating on those areas obviously heavily infested. Make spot treatments only if you can watch the lawn closely day to day.

Recommended materials include Diazinon, Aspon, Baygon, Ethion, and Trithion. When using sprays, it is important to apply the insecticide in a large amount of water. Use one of those quart-size jar attachments to garden hoses that calls for 15 to 20 gallons of water passing through the hose to empty it. The granulated type applied with fertilizer-spreader machines should be washed down into the turf mat after application.

Use your insecticide with careful attention paid to label instructions, and don't expect one treatment to provide insurance against damage for a whole season.

Q. My centipede grass is full of Bermuda. How can I kill the Bermuda?

A. You can't use any chemical to kill Bermuda in centipede. You probably would kill the centipede and perhaps harm the Bermuda very little.

You should have little trouble with Bermuda in centipede if you grow centipede "on the poor side," as it should be grown. Bermuda requires prosperous conditions. Centipede can't stand prosperity. Thus centipede should receive very little fertilizer — if any.

Q. What can I use to get rid of purplish tinge on my grass and to green up some areas that are pale or yellowish?

A. Lawns that are not showing a healthy green should be fertilized with a readily available nitrogen fertilizer.

Materials such as ammonium nitrate, ammonium sulfate, nitrate of soda, or urea are quickly picked up by the grass and will eliminate the purplish tinge and the yellow. They are all high analysis nitrogens and must be used sparingly.

Natural organics such as sewage sludge, cottonseed meal, castor pumace, etc., are slow-release nitrogens of fairly low analysis, and during the cool weather would not give fast results. Because of the filler material in the sludges that tend to lie on top of the ground for long periods, the fungus problems that are usually present during periods of damp, cool weather will tend to be aggravated.

To 1,000 square feet of area, the following amounts of nitrogen should be applied and thoroughly watered in: nitrate of soda, six pounds; ammonium sulfate, five pounds; ammonium nitrate, three pounds, and urea, two pounds.

Ammoniacal nitrogen is usually converted by bacterial action to nitrate nitrogen before it is used by plants but if the soil is saturated with water, too dry or too cold, this process is slow, so nitrate nitrogen is usually favored. A combination of both ammoniacal and nitrate nitrogen, as in ammonium nitrate, is very good.

Q. This year I intend to use a preventative program against chinch bugs as follows: Start spraying in April and spray every two months through October. I have come up with this plan on my own and without professional advice. Is it sound?

A. The Florida Extension Service recommends that South Floridians make their first treatment in a preventative program in March and subsequent treatments at about six-week intervals. In average years, the initial treatment can be made later the farther north a lawn is located, so that in the northern third of the state late May usually would be soon enough to begin. You're not far off.

Q. I would like a little information on the care of Argentine Bahia grass. What type of fertilizer is best? Must the clippings be picked up at every cutting? Can the clippings be used as a mulch or for compost?

I have been told this strain of grass does not decay. Is annual raking advisable?

A. One of the standard grass fertilizers should keep your Bahia lawn in good condition. This grass can be grown with a low level of maintenance but it is not harmed by more frequent fertilizing, as centipede would be. In fact, you'll notice a definite improvement.

It's a good idea to remove the clippings from any lawn. Any plant material is suitable for a compost heap; the finer it is and the more frequently it is watered, the faster it will break down. An annual raking is not necessary.

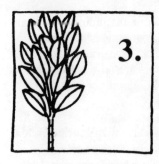

3. Shrubs, Foliage Plants, and Vines

For special index to this chapter, see back of book.

Now you see it, now you don't! That's how fast problems on tropical shrubs and vines can appear and disappear with prompt treatment.

Establishing a small "medicine chest" of remedies helps avoid procrastination. You don't need to mix a gallon at a time or buy expensive spray equipment, either. A hand sprayer and some measuring spoons will take care of most problems on shrubs that can't be hand-picked.

Q. How often should I prune my plants and are there any shrubs or vines that need little or no pruning?

A. Few plants ever "need" pruning to grow well, except for those in containers.

Pruning serves mainly cosmetic purposes — to create more branches, to make a leggy shrub bushier, and, for plants that bloom just on the tips of long branches, to create more tips and hence more flowers.

Pruning can be done at any time of the year in most of Florida, but extensive pruning should be held until spring so that the flush of new growth that results after pruning won't be nipped by cold.

Most pruning of flowering or fruiting shrubs or vines is delayed until after the flowering season, however.

There are many plants that shouldn't be pruned, including cacti and succulents, bamboo, elephant-ears, tree ferns, spathiphyllum and many foliage plants, *Jatropha podagrica,* crinum lilies, bird-of-paradise, and more.

Acalypha hispida, the chenille plant, is one of the most striking ornamental shrubs in South Florida. Its long, fuzzy, red racemes stand out in a hedge of greenery. It is one plant that will flower well without full sun.

The only reason to prune a vine would be to contain it to a certain area. Pruning would not in itself cause a vine to flower more freely.

The best way to avoid bothersome pruning is to plant with foresight by consulting the projected heights and growth rate of plants in books listed in the Bibliography.

Q. The buds on my plants won't drop — they just hang on and won't open. What's the problem here?

A. Bud drop. The term doesn't imply a time limit. Undoubtedly your buds fall off, although they may wait until they're good and ready, which usually means brown and shriveled. A cultural imbalance (too much or too little water or fertilizer) is usually to blame, but thrips and other headaches could be at fault.

Q. Why should the leaves at the bottom of my hedge be few and far between? The top part looks good but I certainly wish I could do something about the scraggly bottom.

A. Maybe it's the way you've pruned it. In developing a shaped hedge it is important to be ruthless in heading back the plants the first year so that branching begins close to the ground. In succeeding years, it's easier to keep foliage on the lowest branches if the top is not allowed to become wider than the bottom. Otherwise, the bottom will be shut off from the sun.

Q. How close to a house should plants or hedges be planted in Florida? I understand that if you get them too close, you'll have a bad mildew problem.

A. Minimum recommendation is 18 inches for small leaved, slower growers. Two feet is the shortest distance recommended for bigger plants. The more space, the better — not only for your walls' protection and maintenance but to promote better air circulation around plants.

Q. I bought a little hybrid plant called 'Many Times' at a hibiscus show. It is only a foot or so high and already has three buds. How should I take care of it?

A. As with other hibiscus, plant it, feed it lightly, keep it watered, and in a year you'll have something uniquely gorgeous. Well drained soil is most important, so avoid planting holes prepared with large amounts of muck and marl. Mix peat moss into the soil you put back in the hole. A once-a-month feeding with a 40 percent organic fertilizer such as 6-6-6 will keep it in good shape. To improve blooming, you might try "bloom-aid" fertilizers with a higher potash content (the third number on the bag).

Q. True or false: you shouldn't let grass grow around hibiscus?
A. Ideally, it's better not to let grass creep up on them. Hibiscus don't like too much competition for moisture and plant food from grass or weeds. If you're not religious about midsummer, early fall and late winter feedings, then it's best to trim back.

Q. I've heard that grenade scale is a problem with hibiscus in South Florida and this has kept me from playing around with them. How serious is it?
A. Members of Florida hibiscus clubs aren't discouraged by the scale. They spray routinely for it with Cygon at two-week intervals in hot weather, and say it seems to trouble old or rundown hedges most. Telltale signs are shedding of flowers, buds, and yellow leaves.

Hibiscus should be examined weekly for scale, especially in the summer. The round black or brown bumps found mainly on the stems may be rubbed off with a twig or sprayed with Cygon if there are several plants generally infested. Pruning back first and fertilizing afterward may be desirable if scale is well established.

Q. I discovered that my scarlet hibiscus hedge was losing its leaves and dying back because of something called grenade scale. Some of my neighbors have the same problem. How bad is this and do I have to cut back my hedge to stumps or just spray?

A. Grenade scale, also called hand-grenade scale, hibiscus scale, and hibiscus hand-grenade scale, was first spotted by Dade County Agent Lou Daigle and R. I. McMillan, formerly of the state Division of Plant Industry, in 1962 on an acalypha and a hibiscus. Since then it has given hibiscus lovers a headache, spreading to Broward, Palm Beach, Lee, Sarasota, Manatee, Pinellas, Brevard, and Volusia counties.

It is said to be most active in Volusia but continues to play havoc with many Dade hedges. In addition to the 'Snow Queen' and 'Formosa' hibiscus and the rose-of-Sharon, its favorite hosts are the copper-leaf (acalypha), ragweed, spindle-tree *(Euonymus japonicus l.)*, governor's plum, gardenia, geranium, Barbados cherry, turkscap, and pittosporum.

The sucking pest prefers hibiscus, though, and can only be controlled if prompt action is taken. Hedges should be examined closely for small yellowish-brown bumps on the stems. Infected plants should be cut back severely, all clippings destroyed, and the plants sprayed at least three times at four-week intervals with Dimethoate (Cygon 23.4 per cent or DeFend 267EC) at one cup to 25 gallons. Severe infestations may require additional spray treatments. (Be sure to follow safety precautions on label.)

Grenade scale is found in Cuba, Panama, Puerto Rico, and the Virgin Islands as well as Florida.

Q. We transplanted a double yellow hibiscus and now it has several healthy looking buds on it. Just before the bud is ready to open they seem to be snapped off about an inch or two from the bud. The plant looks very healthy and we don't understand what is causing this. There are no bugs on it, and a hibiscus nearby is blooming nicely.

A. A bug isn't chomping off your buds; it was your shovel. Transplanting can usually be expected to interfere with blooming on many plants and trees. It disrupts the flower-forming process so that the plant can only take bud development so far. Then there's that sickening thud as what looks like a perfectly good bud snaps off its stem and hits the ground. It's a temporary condition, but that doesn't make it less disappointing.

Q. Can you give me a fast course on hybridizing hibiscus? I understand that if you do produce a brand-new-looking flower you can't propagate it by seed. How come?

A. The fact that you can't put hibiscus hybridizing in a nutshell is what makes it such a fascinating hobby. Basically, you take pollen from the anther of one variety and place it on the stigma (the "pom poms" at the top of the pistil) of another. Then you harvest the seeds, plant them, and in from 15 months to two years, your seedlings will present you with blooms — all different and maybe even one an extraordinary surprise.

Meanwhile, if you're like most hibiscus enthusiasts you'll be collecting other people's hybridized flowers. Since the seeds do not run true to the flowers, the only way to propagate winners is by grafting them onto the hardy root stock of old favorites that have proven successful in your soil type. 'Anderson's Pink' is a good root stock for this area and so are 'Brilliant' and 'President' for those who have trouble with nematodes. 'John Paul Jones' is a popular choice for double varieties.

Q. "Black splash" sounds messy — what is it? I've heard my gardener mention it a few times but as he doesn't speak English very well I can't seem to pin it down.

A. Black splash isn't a new wall paint, as the name might suggest, but a mysterious plague on hibiscus. Symptoms are elongated black spots on young stems and irregular spots on the leaves. Lower leaves are usually affected first. Later the leaves turn yellow and fall, although some areas of the fallen leaves may remain green.

Other symptoms include bark sloughing from the plant and roots, death of the new shoots, and decline and death to the plant. Although the exact cause isn't known yet, pathologists suspect that black splash could result from poor nutrition, waterlogged soil, or heavy infestation of root-knot nematodes.

Until the case is known, try pruning out the ailing wood and improving the growing conditions.

Q. My husband's the gardener in our family and I have to watch him mope around for days when his remedies for plant problems don't work. He's currently grouchy about the number of buds dropping off our hibiscus. He's ruled out nematodes, aphids, and thrips. What else could be the matter?

A. Poor drainage or shock from recent transplanting are two possible causes. But the most common is a deficiency of one or more plant food elements. Hibiscus should be fed in midsummer, early fall, and late winter with a fertilizer containing from four to eight percent nitrogen, six to eight percent phosphorus, and four to eight percent potash. (Those percentages are the three numbers listed on the bag.) A small plant should receive about three tablespoonsful and a larger bush about two quarts per application.

Q. Why do some roses do well for me and others do poorly, even though they are planted in the same bed? Some have very sturdy stems while others become limp and do not stand up when a flower opens on the stem.

A. Acclimatized rootstocks are vital with roses. Stay away from supermarket "buys" in fresh, bare-root arrivals from out-of-state.

Roses grafted on *Rosa fortuniana* rootstock (double white 'Cherokee' or 'Evergreen Cherokee') grow larger, more vigorously, produce more flowers, and live several years longer than plants grown on any other rootstock, veteran growers have determined. 'Dr. Huey' ('Shafter') is second best, and 'Multiflora' *(Rosa multiflora)* brings up the rear as the shortest lived here. Most nurseries in South Florida sell these grafted types, but ask to make sure.

Q. Where and how should you plant roses here?

A. Roses need a location that gets direct sun for at least six hours a day. Morning light is desirable for drying dew on leaves and reducing the chances for black-spot infection. Pines make nice neighbors because their tap roots forage for nutrition well beyond the range of rose roots, which don't like competition from other plants.

Roses like organics and should have a bed made from equal parts of composted cow manure, peat moss, and nursery topsoil. The planting hole should be 18 inches deep and a foot wide. If you are unaware of the situation in your soil regarding crown gall organisms and nematodes, it's safest to fumigate with Nemagon or Vapam two weeks prior to planting.

Q. My rose plants get spotted, the leaves turn yellow and fall. The leaves also become twisted. What's the problem?

A. First, black spot and, second, powdery mildew, a fungus which curls leaves and affects buds and stems.

For continually healthy plants, these two problems must be control preventatively. Many growers spray weekly, alternating Benomyl with Dithane M-45 or one of the many other fungicides available. These sprays also will control another rose problem, cercospora leaf spot, recognized by its distinct gray centered spots with purplish or reddish brown borders. This problem usually is associated with neglect or poor drainage.

Q. I water my roses every day, spray them once a week, and fertilize them once a week, and yet the leaves turn brown and fall off. What am I doing wrong?

A. There should be no need to water roses daily. A good soaking two or three times a week should be plenty. When you water every day, there is a temptation to put down a little bit of water each time. Most of this is lost through evaporation, and the root zone may be bone-dry.

Apply the water to the soil rather than to the plants. A slow-running hose will soak the root zone. Avoid keeping the foliage wet for long periods by not spraying the leaves.

Rose plants can be burned by chemical sprays or fertilizers if they are not used correctly. It's a good idea to fertilize or spray plants the day after you have soaked the root zone.

Spraying to prevent or control insects and disease should be done weekly or a little more often in periods of heavy rainfall. Get good coverage of the plant, including bottoms of leaves, or much of your effort may be wasted.

There's no need to fertilize roses weekly. A 6-6-6 fertilizer applied monthly or every 10 weeks should keep the plants in good condition.

Q. I've heard that you can add various things to the soil to make your roses a deeper color. Iron, for example. Before I make a trip to the nursery, what's the story?
A. No one element will do it. Other factors govern the intensity of flower color. The more foliage on the plant, the brighter and more intense the coloring. And in hot weather, the colors are less intense because food from which the pigment is manufactured is used up in respiration. Bright sun and premature leaf fall caused by disease also lessen color intensity.

Q. I want to cut back my scrawny roses, but I understand June isn't the time to do severe pruning. How severe should I be?
A. In November and March, roses should have tall, leggy growth removed — some of the extra-slender canes in the center of the bushes. Be sure when you are pruning roses, or cutting them for the table, to use only sharp tools since breaking or twisting stems injures the remaining wood. Avoid cutting any longer a stem than necessary. The more leaves you retain, the better able a plant is to manufacture its own food.

Q. I'm stuck on what to do for my Spanish bayonet (yucca.) One of them has brown, oval, sunken spots on its leaves, another has gray spots edged with brown and sick-looking leaf tips.
A. Control measures for the fungi that cause this spotting and the tip necrosis haven't been developed, so it is recommended that infected leaves be removed and destroyed. To help make sure they won't reappear, spray your plants regularly with fungicides like Daconil, Maneb, or Zineb.

Q. When moving to Miami last year, I brought my Christmas cactus

with me. It has made fine growth, but last fall did not bloom. Is there any particular treatment or growing conditions I can give it through the coming summer months so that it will bloom this fall?

A. The plant should have partial shade, moderate watering, and a light application of fertilizer monthly during the summer. Best light is that found under high trees like oaks.

Just make sure your plant doesn't receive extra light from a porch or street light during the season when days are naturally shorter. The plants need this reduction in light to bloom.

The Christmas cactus is included in a group known as crab cacti. There's a Thanksgiving cactus, and an Easter cactus as well.

Q. I have a large, fruit-bearing cactus on a patio that is protected from rain. Recently a fine, white, powdery scale has appeared on all the branches of the cactus. Is this because of lack of rain, or does it require a spray?

A. Without seeing the plant, a safe guess would be that it is a cactus scale which could account for the powdery appearance on the plant. A Malathion spray would control this. If such a spray is used, wait a couple of weeks before using any fruit. Follow directions and warnings on the label.

Q. What causes the leaves on cactus plants to split and rot? Do they need much water?

A. Most of the trouble with cactus plants comes from overwatering. Rotting is due to excess water. Don't water more than once every 10 days to two weeks. Best rule to follow is to keep the soil slightly moist, never wet.

Q. How do I know when the fruit of night-blooming cereus is ready to eat?

A. The fruit is red, oval-shaped, and about three inches in diameter. It does not often mature on this species in Florida, however. When it becomes dark red and soft, it is ready to eat.

Q. I have a lot of night-blooming cacti and lately the plants have become very gray. Some of them are even rotting out. What's the trouble and is there anything I can spray on them?

A. Cacti here are subject to this deterioration by fungi which build up in moist warm weather. Best procedure is to cut out the affected areas and spray with Maneb, Zineb, or neutral copper.

Q. I've never had much luck with cactus here although I can't resist

buying them. They'll do fine for a year or so, then seem to wither away. On some, but not all, of these sick plants I've found tiny, football-shaped bugs that are gray and sort of fuzzy. How can I get rid of these and what do you suppose is the matter with the plants that die but aren't infested with these bugs?

A. The insects are mealybugs, and they and other kinds of scale may bother cactus. You can control them with Diazinon. Mites are sometimes a problem, too, and they can be curbed with Kelthane or other miticides.

Rot is something that creeps up on many cacti in this area, usually following injury or excessive moisture in the air or soil. When it is first noticed, the infected tissue should be cut away. In advanced stages of rot it is necessary to remove all of the infected area to the point where the plant may lose its value as an ornamental. Poor drainage can be a factor in encouraging this condition, as well as too-deep planting.

Cacti should not be watered more than once every two or three weeks during the growing and flowering season — much less often during the cool months of the year.

Q. What can I do to make crotons grow faster? Should I cut them back?

A. Make sure drainage is adequate. Crotons tolerate many kinds of soil. Usually they need little more than spraying for insect control and pruning for shape. I think a little 6-6-6 fertilizer would be fine. You might try a water solution of sodium nitrate at the rate of one level teaspoon to the gallon. Apply once a month. A little cow manure would also be good.

Q. What can make a croton leaf lose its color? One of my plants is turning a very insipid light green, and the spots are fading from bright yellow to cream.

A. Could be those tiny sucking insects with the undignified name — thrips. The thrip is pale yellow when young and dark brown or blackish when mature. By draining out the plant juices it brings about the color change in leaves. You can control thrips with Malathion, Meta-Systox-R, or dimethoate (Cygon).

Q. My crotons are too narrow and stringy for my taste, but I don't want to reduce their height by pruning them to force a more plump shape. Is there any way I can fatten them up by feeding them or something?

A. You might try girdling them, which sounds like you'd be making them look even slimmer. But it has the opposite effect. By taking a knife and removing a ring of bark about a third- or a quarter-inch wide, you'll encourage new branches to sprout below this cut on every branch you try it. Done carefully, the branches above the cut will remain healthy. Experiment on a few branches first.

Q. I am having some trouble with my crotons. They get little discolored spots all around the edges, and eventually these widen into big brown spots. Finally the whole leaf dies. What do I do?
A. Your crotons have a leaf-spot fungus that neutral copper will correct. A series of three sprays at two-week intervals should clear it up.

Q. Leaves from my hybrid crotons are dropping off. Spraying for red spider has not helped. I give them about two handfuls of 6-6-6 fertilizer about once a month. What is wrong?
A. Assuming you used an appropriate spray for spider mites, experts suggest an application of Malathion for thrips, which also cause leaves to become dry and fall. Don't overdo the fertilizer. Every two or three months would be better.

Q. I've been told my yellow ixoras have an iron deficiency problem (the leaves are turning yellow and the veins are a deep green). I'm for burying some nails in the ground but a friend says it should be sprayed on. I have plenty of old nails but before I spend money on spray I'd like to know what I'm doing.
A. Toss out those nails; iron rust isn't an available source of iron to your plants in alkaline soils. And spraying with iron is the hard way. Apply chelated iron to your soil for best results, but be sure to water well the night before and flood it in with water. (Don't put this on the leaves.)

Q. My ixora have developed a black fungus on many of the leaves. What causes this and what should I do?
A. The small, translucent, green shield scale has been causing havoc on many shrubs and fruit trees. Its victims include gardenias, ixora, citrus, and other fruit trees.
The almost-invisible scale secretes a transparent, sticky, honey-dew material on which the black sooty mold develops. Control of the scale is Malathion or dimethoate (Cygon) as directed on the container.

Q. Our 'Super King' ixora are strong and flourishing but do not bloom. There are 10 plants against the south side of the house in full sun. Two months ago I asked the advice of the nurseryman from whom I bought them, and he gave me a mixture which only enhanced the

problem. They're growing faster than ever, and there is only one bud near the bottom of a plant. What could be wrong?

A. It's possible to keep pushing a plant with fertilizer and water with the result that there is plenty of new growth and foliage, but no bloom.

Where bloom is the objective, you will want to use a fertilizer with more potash than nitrogen. A 5-10-10 fertilizer or one similar to it will do. This should be an acid-forming fertilizer for ixora.

Generally, a moderate course is a good bet. If your fertilizer program has been heavy, slack up a little. Ixoras like to be fairly moist, but you don't need to overwater, either.

Q. I was determined not to have just another red ixora hedge, so I put in a row of those beautiful pink ixoras. I didn't have any problems with them the first year but this year they dropped their flowers despite good care. What could be the matter?

A. It may not be your fault. The bud drop could be due to (1) too much humidity, (2) too little humidity, (3) too much soil moisture, (4) too little soil moisture, (5) sustained high temperatures, or (6) strong, blustery winds. If your bushes are not planted where one of these conditions exists permanently, you'll probably have healthy plants after they recover.

Q. Remind me again, please, of the time to prune poinsettias.

A. Approximately March, June, and September. If you pruned your plants back after the blooms faded and cut back each time the new growth reached a length of 12 inches, your plants should be a pleasingly plump shape by now.

Depending on the size you want your plant to be ultimately, you can prune your plant back as much or as little as you want. Regardless of the length you cut back, it's the number of times that matters since two or three branches will appear at each cut. Flowers (bracts — modified leaves — actually are the colorful parts) appear only at the tips of each branch.

Pruning after September 10 will interfere with the setting of the buds, which happens soon after October 10 when the days shorten.

Q. My poinsettia went limp a few weeks ago for no apparent reason. Since then I've kept it well watered, but, if anything, it looks worse. Nothing seems to be affecting the leaves. I can't see any bugs or bumps. What could cause this?

A. Plant wilt often shows up after periods of very heavy rainfall. This condition is more likely to be noticed on tender succulent plants such as poinsettias, avocados, papayas, and hibiscus. It's caused by long-term root contact with excessive water which suffocates or drowns the tender feeder roots.

When the plant is subsequently exposed to periods of dry weather, it will quickly wilt, drop many leaves, and possibly die. The best course to follow is to fertilize lightly and frequently to encourage rapid new root development. A light pruning is often suggested to decrease young tender leaves and stems to compensate for damage to the root system.

Q. Leaves of my poinsettia plants are dropping. What is the cause and the cure?

A. The chief insect pest of poinsettias is a large, green hornworm. Best way to get rid of it is to pick it off the plants and kill it. Spraying with Sevin may also be helpful. Poinsettia scab also attacks these plants and is hard to control. It is usually best to prune all infected parts and destroy them. But since it is so close to the flowering period, why don't you try spraying with a neutral copper fungicide or Ferbam? Repeated sprays may be necessary.

Q. My gorgeous double poinsettia is dropping its leaves and the new ones are twisted. I suspect poinsettia scab, a problem my neighbor had. She lost hers — how can I hold onto mine?

A. Identified by raised lesions on the stems, the scab is a fungus readily spread by spores in the summer through wind and rains. Infected areas should be removed from the plants and burned or wrapped and placed in trash containers. Don't put the cuttings on the ground near the plants while pruning; deposit them on a paper or in a container.

Then spray weekly with Maneb, Captan, or Thiram. Some people, as a preventive measure, use these sprays every two weeks to thwart development of the fungus.

Q. My friends have insisted that I write about my rather unique poinsettia plants. I put several poinsettia cuttings in the ground, and instead of sprouting leaves, they have tiny poinsettia blooms that look pasted on up and down the stalks. There are no stem or leaves. How did this happen?

A. The experts believe this odd situation can be blamed on the weather. Unfavorable for growth, it has not prompted the conditions which would push your cuttings out of their reproductive cycle and into vegetative growth. So they are retarded in the state in which you removed them from the mother plant.

Q. What pests should I watch for on my poinsettias?

A. Mites (very, very tiny spiderlike creatures) and caterpillars are sometimes a nuisance. You can treat the mites with Kelthane or Tedion. Caterpillars are usually few in number and can be knocked into a container for disposal. Or you can spray with Sevin.

Q. After seeing (and sniffing!) my neighbor's gorgeous gardenias, I'm going shopping. What's the best variety for South Florida?

A. To begin, choose grafted plants — maybe an inexpensive young 'Veitchi,' a 'Radicans,' or a voluptuous and expensive-looking 'Miami Supreme.' All of these are a form of *Gardenia jasminoides*, yet very few dealers can differentiate between the varieties although each differs in form and size of leaf and flower. A good growing medium is a mixture of five parts marl, one part sawdust or peat moss, and one part sand.

Q. I want a gardenia hedge along both sides of our patio and see no reason why I can't use the cuttings I have taken from the large plant by our carport. A neighbor as much as says that I am wasting my time and should buy small plants from a nursery. Why? I have a friend who has grown three husky plants from cuttings.

A. Facts back up your neighbor. Most of the enemies of gardenias can be controlled but the worst of these is the rootknot nematode. When grafted to *Gardenia thunbergia*, gardenias can tolerate nematodes although they're not 100 percent immune. You could still graft your own but with low-cost grafted gardenias available, this is an exercise only for those who enjoy grafting. Of course, potted (in sterile soil) gardenias would have a chance.

Q. Is there an all-purpose spray I can use in controlling insect problems on my 'Ameii Yoshioka'?

A. Several insects, including whitefly and scale, damage gardenias by withdrawing juices. Sooty mold often follows a buildup of these insects. Isotox Garden Spray (sold in all garden stores) will control most; mix Ortho Volck Oil Spray (but not in warm or cold weather) with Isotox to kill scale. Be sure to get undersides of leaves and never spray a weak or wilted plant. Repeat in about two weeks.

Q. What causes buds to drop on gardenias? I hear this is a common problem, but what's the cure?

A. You might blame it on excessive or unseasonal rainfall, although it's usually the fault of the gardener. Poor watering, poorly drained soils, excessive fertilizing, not enough sun, and mechanical injury often prompt bud drop. Root injury as a result of nematode infestation and fungus root rot are two other possibilities.

Heavy fertilizing should be avoided; frequent light applications are best. Over-fertilizing over a prolonged period or in one application can cause dehydration of root systems because of the accumulation of excessive soluable salts. After it has rained heavily, hold off watering for several days to let the roots dry out somewhat.

Q. I understand gardenia bud drop is caused by some sort of imbalance in growing conditions. What conditions are best for good blooming?
A. The secret of vigorous, free-flowering, disease-free plants is unchecked growth. Full exposure to sun the year-round, sufficient humidity, constant soil moisture, and rich, quick-draining soil are necessary for almost all varieties.

Q. Here's a mangy leaf from my poor gardenia plant, and I'd like a diagnosis. The whole bush is sad-looking — maybe because it, along with the rest of the back yard, sat in two to four inches of water recently.
A. Your leaf shows quite a collection of tiny soft scale insects. Those colored orange are acuminate scale. Those that look like tomato seeds bordered with white are pyriform scale. The yellow splotches could be remnants of the so-called friendly fungus (usually less active in the fall-winter and, at any rate, not harmful). The other two, however, should be controlled with a couple of sprayings with Cygon or Ethion.

Q. What is the "friendly fungus?" I pointed out some big white spots on her gardenias to a friend of mine and that's what she called them. She said her nurseryman told her they don't do any harm.
A. "Friendly" probably is a species of aschersonia that attacks the nymphs of the cloudy-winged whitefly — which isn't so friendly. Sometimes mistaken for a scale insect, it can grow as large as a quarter-inch in diameter.

No control is needed for this beneficial fungus, but its presence indicates that if it can't handle the job, the white-fly larva feeding on the underside of the leaves should be sprayed with Malathion or oil emulsion.

Q. Despite the fact that they're not supposed to grow well in South Florida, I've seen several husky azalea bushes in the area, all loaded with blooms. But their owners must know something I don't. What's the secret?
A. Azaleas take a while to become good producers here. They like a soil mixed with plenty of peat moss, a good oak leaf mulch, and regular

watering. Your soil should be remodeled so it is less alkaline by adding manures in generous amounts, and a gardenia-type fertilizer applied a couple of times a year helps, too. Plant your azalea at the same level it was in the can; too deep planting causes problems.

Azaleas aren't salt tolerant and they don't like grass infringing on their territory. Nutritional sprays should be applied when leaves show yellowing. These problems plus thrips, mites, and a flower-spot discourage many gardeners from naturalizing the plants here. But, of course, there are always exceptional plants that look good year after year.

Q. Some of the leaves on my crape myrtle are a sickly grayish white and twisted, and some of the flower buds don't look too good. What's the matter?

A. Most troublesome in the fall and spring, powdery mildew can still be a problem on crape myrtle. The fungus attacks young, tender growth, thrives on hot, humid days, and is scattered by tiny seeds or spores that float on the air currents until they reach your tree. Safest controls are Karathane and ActiDione PM. Sulphur and Phaltan may be used when the temperature drops below 80 degrees.

Crape myrtles are also prone to a magnesium deficiency, which shows up in a change of leaf color from green to a reddish brown along the margins and tips. Correct it by applying Epsom salts or Emjo to the soil.

After your crape myrtle blooms, cut it back about a foot from the blooming tip and you may be rewarded with a second blossoming later on.

Q. When we bought our house I was told by several people that one of the shrubs was a Brazilian pepper and that it would have large clusters of red berries in the winter. In two years it has produced tiny blossoms but no berries. Does is require special feeding?

A. No. Brazilian peppers — often called Florida holly — need no special attention or soil, but to insure their fruiting it is necessary to have both male and female shrubs since they are unisexual.

This is the number one pest plant in South Florida. Although it can be shaped up into a nice bird-food tree, it grows very fast and its dropped berries sprout all over the ground under the tree.

Q. I would like to know what to spray on oleander for caterpillars.

A. The pest quite likely is the larva form of the oleander moth. This caterpillar is orange in color and has tufts of long black hair. A chlordane spray is suggested, but the spray may have to be repeated rather frequently for control. Follow all directions and warnings on the label.

Q My oleanders have developed stringy twigs riddled with pitted bumps. Does this sound familiar, and what can I do about it?

A. It sounds like pustule scale, a tiny armored scale first discovered in Florida in Orange County in 1942. Since then, it has spread, not only around the state but to other ornamentals.

The controls are dimethoate (Cygon) 25 percent EC, or Diazinon 25 percent EC, or Malathion 57 percent EC — each used at the rate of a quarter pint per twelve-and-one-half gallons of water.

Q. In January I bought a camelia plant and by March there were no leaves left. I replaced the plant about a month ago and now the leaves on this one are turning brown and falling off. Any suggestions?

A. There are several possibilities for you to investigate. Your plant may be in an area of poor drainage. Transplant it now, or raise the plant if it has been set too deeply in the soil. Camelias don't do as well in South Florida as in some other areas of the state. They are very sensitive to lime in the soil, especially in sandy soils. So, you must provide an acid soil preferably between pH 5 and 6. Put at least 75 percent straight peat moss in the planting hole and plenty of organic material. It will do best for you in partial shade.

Q. I was given a little bird-of-paradise plant two years ago, and although it appears to be in good shape, it hasn't bloomed. Maybe I'm doing something wrong. What conditions does it like best?

A. *Strelitzia reginae* need a good organic soil, broken shade, a lot of water — and at least three or four years to mature to the flower-producing stage, which may be your problem.

Q. My podocarpus seems to be healthy since it produces a heavy crop of fruit, but the leaves are tipped with yellow. I have two questions: Can I eat the fruit, and can I get rid of the yellowing?

A. Pistillate *P. nagi,* which bears heavy crops, is known to lean toward yellow-tipped leaves. Female podocarpus are the only seed-bearers and should have additional magnesium applied in the spring. The purplish flesh above the seed is edible and can be used for jellies, but it takes reeducated tastebuds, in most cases, to appreciate the flavor.

Q. I have a hedge of *Carissa grandiflora.* Occasionally one of the plants drops its leaves and dies. Plants next to sick ones continue to thrive. Do you think the blight is being caused by nematodes around the roots?

A. The problem probably is not nematodes. If it were, you would see a gradual reduction of health and vigor over a three-to-four-year period.

A fungus disease could be the problem, particularly if you notice galls (enlarged areas) on the stems. Severe pruning of affected areas and spraying with Fore or Benomyl — perhaps even on a twice-monthly schedule until new, healthy growth is established — will help.

Q. We have a datura in the planting on one side of our patio and for three years it has not bloomed, although it looks healthy. I have followed suggestions of friends who have plants that bloom well, but nothing seems to work for me. Have you any recommendations?

A. Perhaps your datura is crowded by your other shrubs. Daturas like plenty of space to stretch their roots. They also need a well drained soil despite the fact that they like generous watering.

Q. Kindly tell me why I can't grow the gigantic elephant ears that other people do. They come out beautifully but grow only about two feet, then turn yellow and drop off or wither. I have tried iron and more water.

A. Try five pounds of Milorganite per plant and more water. Elephant-ear alocasias are very thirsty plants, and you'll be surprised how fast they'll respond. Even a daily soaking isn't too often, especially in a fairly sunny spot.

Q. What do you do for spotty crown-of-thorns plants? I'm enclosing a sample.

A. Plants in a weakened condition are susceptible to attacks by these leaf spot fungi and it's best to restore them to good health before applying fungicides. By then you probably won't need to use a control because healthy new growth will have appeared. A complete fertilizer such as 6-6-6 or 4-8-8 will help new leaves develop quickly.

Q. I left my two potted rubber plants in the bathtub when we went away for two weeks. When we came back the tub was covered with yellow leaves and the only signs of life on the plants are the green tips. Do these green tips mean the plants will live or did we make a fatal mistake?

A. A mistake — but very likely not fatal since ficus are hard to "do in." Over-watering is one of the prime causes of yellowing and dropping leaves. Other factors can be air pollution, low light density, chilling, poor soil drainage and aeration, and root decay from soil-borne diseases or insect pests.

People usually over-guess the amount of water their potted plants need.

If you are going to be away an extended length of time it is better to fill a bucket with gravel, half-fill it with water, and run wicks from the water into the pot.

Q. Some years ago we planted bamboo (in our ignorance) at a house we have owned but kept rented. We have returned to this house now to find the bamboo running wild. It has made a good screen but it is coming up 50 feet away in a neighbor's yard.

Can you give us any suggestions as to how to go about keeping it under control in just the area in which we want it?

A. Running bamboo spreads rapidly. Its growth must be restricted or it will soon form a thick jungle that extends many feet beyond the original planting.

The USDA Agricultural Research Service says a curb made of sheet metal, concrete, or asbestos board will prevent bamboo from spreading. The curb must surround the planting. The top of the curb should be about one inch above the soil surface and the curb should extend several inches into the ground. Any joints in the curb must be lapped and secured tightly; bamboo rhizomes can force their way through very small openings.

Buildings, wide driveways, and roads also restrict the spread of bamboo. If the bamboo canes come up in a lawn area, mowing will destroy unwanted canes by cutting them while they are still small and soft.

If you plan to grow bamboo without curbs, be careful in choosing a planting site; protect yourself and your neighbors from unwanted bamboo in flower beds, hedges, and shrubbery plantings.

To eradicate an entire planting of running bamboo, cut down all the canes. Repeat the cutting. When new canes that follow have reached their full height, cut them all down and the rhizomes will soon die.

Q. What would cause a sago palm trunk to develop a large crevice running perpendicular almost the full length of the trunk? For a depth of several inches the outer "shell" has pulled away from the core of the trunk, and between this outer bark (this pulls off easily in some areas) and the core there is a rather mushy dark residue.

I was told that this can happen to sagos over 10 years old and the only thing to do is to let the offshoots grow up to cover the dead stump. Should I give up?

A. The split could have started in one of two ways. A mechanical injury may have opened the plant to a fungus invasion or stress caused by excessive moisture could have prompted the fissure, opening the way for fungus to center. An attempt to rescue the plant would include removing all loose bark, cleaning out all the "mushy" material, and thickly dusting with a powdered fungicide. After it (hopefully) dries up, the split should be treated with a wound asphalt paint, available at garden supply shops.

By the time a lot of gardeners notice a problem such as this it's often too late to do anything. It may be more practical to transfer your affections to the plant's offsets or a new cycad.

Q. I brought two bonsai here with me from up North where they were thriving nicely. One is an evergreen, the other a pine, and both are now turning brown.

I have certainly given them the care they require — which is considerable — but I can see them dying daily. I'm anxiously looking for advice.

A. Bonsai — ordinary trees made miniature by systematic root and branch pruning — haven't lost any of their individual cultural preferences, despite the fact that their appetites are mini compared to their full grown sisters. A bonsai white cedar is going to want the same climatic conditions as a naturally grown cedar. If you bring this tree to Miami from a colder, drier climate it will react to the change negatively. Even though you cut down on the amount of water you give it to compensate for the humidity, it is likely to be more prone to attack by tropical pests and diseases than trees native to this climate zone. These warm climate trees are the ones chosen by local bonsai fanciers for successful growing here.

Q. I bought a chenille plant recently and would like some more information on it.

A. The chenille plant is an acalypha, a relative of the copperleaf, and a fast grower, reaching a height of about eight feet and a spread of six feet. Effective as a specimen, border, or container plant, it dangles its odd red tails in the great profusion in warmer months but flowers periodically throughout the year.

Acalypha hispida likes full sun or partial shade and a well drained soil. It propagates easily by cuttings and should be pruned after flowering to check growth. Scales, mites, and aphids are its occasional pests.

Q. I'd like to put in a bed of dieffenbachia around a corner of my house. Half of it will be in shade most of the time and half in sun. Would dieffenbachias take both of these conditions and, if so, what care do they need?

A. Dieffenbachias, of course, rose to fame as shade (indoor) plants but they'll grow in the sun, too. The best soil has peat added and leans strongly toward an acid reaction. Fertilizer should be applied lightly once a month during the warm months, and the plants should be watered moderately.

Malathion will control the mealy bugs, mites, and scale that might appear. Nematodes are best avoided by planting in "clean" (nematode-free) soil.

Dieffenbachias are actually tolerant of many soil types and reactions. To propagate them, lay a whole cane horizontally in a cutting bench of sand and cover an inch deep. New shoots and roots will arise down its length, and the cane can be chopped into many new plants.

Q. What's the best way to root morning glory cuttings? Some say sand and others, vermiculite. How about just plain water?

A. Many plants will root in plain water, or water with a little nutrient added. House plants like philodendrons will put out roots in water but they prefer builder's sand, vermiculate, or perlite. Morning glories and oleander may also be started in water, and if it's direct from the sky, rather than heavily chlorinated, they will really perform.

Q. We have just removed two large hibiscus that extended beyond the property line close to the outside wall of our garage. The now-bare wall is unattractive, but the planting space is narrow and we don't want to replant with vines that will need constant pruning or tying. What would you suggest?

A. An espaliered plant could make a striking effect in the minimum space you have. A pyracantha, for instance, would give you flat, evergreen foliage plus hollylike berries, and it would still be easy to control its growth. A number of other shrubs can be espaliered. After you shop for ideas you may want to experiment with this classic form of gardening, yourself.

Ficus pumila (sometimes called *Ficus repens* in nurseries) is a creeping vine whose leaves will cling to a wall without support. It is ideal for very narrow spots.

Q. Until this summer I was pretty proud of the fact that I had a good-sized bed of English ivy growing under the eaves in front of our house. Now many of the leaves have developed nasty-looking spots with greenish brown margins and reddish brown and black centers. Some of them have halos around them. What is this and how can I get rid of it?

A. As its name suggests, English ivy tends to be more at home a few degrees to the north, although it can adapt to a greater or lesser degree here. Under warm, humid conditions its health can become impared by a bacterial leaf spot such as you describe. The bacterium is spread by splashing rain or overhead water, possibly, in your case, dripping from the eaves.

All infected foliage should be removed and destroyed. Greenhouse growers spray with a material called Agristrep to prevent spread of the disease. In the future, you should avoid wetting the foliage as much as possible.

Q. Could you please let me know how to take care of bougainvilleas that we have just planted? How much water? How to fertilize and spray?

A. Until the plants are well established in the soil, you should plan to irrigate frequently and fertilize lightly each month with garden fertilizer.

When the plants are doing well in the ground, you ought to fertilize about four times a year. During the spring and summer, you can use a 2-8-10 fertilizer. In fall and winter, nitrate of potash is suggested. The plants should be kept on the dry side during the fall and winter after they are established or they will not flower well.

A small, green worm that chews the foliage is apt to be the most common pest. You can use Sevin for control.

Q. I've got a problem on my bougainvillea that I can't identify. There are silvery "tracks" across the leaves that eventually turn black, but I can't locate the pest that makes them.

A. Those "tracks" most likely are caused by the tiny blister leaf miner who raises a very thin layer of leaf tissue as he burrows his way across its subsurface. Spraying with Malathion or Diazinon will help control him.

Q. Can you tell me why a bougainvillea, bought while in bloom, has never bloomed since?

A. Perhaps you are using too much fertilizer, causing vegetative flush. Or your plant may not be getting enough sun. Sometimes transplanting a plant from one location to another will induce it to bloom.

Q. What kind of care does a hoya vine like?

A. These very attractive "wax plants" with their large, showy clusters of white and pink flowers need little water in the winter, more water and light in the summer, and monthly fertilizing in hot weather. Since flowers come from old flower spurs, do not cut these off if you take blossoms indoors.

Hoyas like high, broken shade, a peat-augmented soil, and a watchful eye for nematodes. Mealy bugs can be controlled with Malathion.

Q. A lot of people leave wood roses in vases at the cemetery where I work, and in the spot where we dispose of old arrangements seeds from these pods have grown into enormous vines. I'm crazy about vines

growing wild and am wondering if I just took some of these seeds and scattered them in what is sort of a moist, shady jungle behind my house if they'd grow.

A. The wood rose *(Ipomoea tuberosa),* also known in other areas as the Ceylon morning glory, grows fast from seed in full sun — which doesn't necessarily mean it has to be started in sun, but it will seek light.

Some people scratch the outer covering of hard seeds like morning glories and soak them overnight before planting, but this is insurance an aggressive vine like the wood rose probably doesn't need. Toss some seed around the edges of your jungle and see what happens.

Sidelights on wood roses: the large tuber is edible; flowers appear in the fall and the popular rose-shaped seed pods in springlike weather.

Q. Our oak tree is hung with a sort of vine that looks like a mass of orange roots. What is this and what can we use to get rid of it?

A. There are two types of vines that exhibit this parasitic, take-over habit. Yours is probably the woe-vine, *Casytha filiformis* L., most common on the east coast of Florida. Trees and shrubs infested with the plant look as though they were draped with a yellowish-brown wire.

A perennial living from year to year, it can be controlled by hand pulling if the infestation is light or by cutting it loose with long-handled shears. With heavy infestations, cut the tree back and dispose of limbs strangled in vine.

The dodder vine, *Cuscuta americana* L., looks similar but is a yellow annual. This leafless plant drops seeds from small flowers, and as soon as the young shoot reaches an acceptable host plant, the root dies and the plant becomes parasitic. If it can't find a host, it dies.

Keeping the soil under trees clean is one way to prevent reinfestation. Any spray strong enough to kill these vines would hurt your tree.

Q. What should I do about the leaves on my allamanda? They're light green to yellow and the veins are dark green.

A. One of the elements allamandas are most commonly deficient in is manganese. This is sometimes confused with iron deficiency but in the latter case, the veins are very narrow and light green, and the remedy is chelated iron applied to the ground. For manganese problems, apply a foliar spray containing this element to the leaves.

Q. What's the matter with a passion vine that drops its buds? It's very upsetting.

A. A soil deficiency is usually the problem. Use a fertilizer that contains a good amount of magnesium — at least half the amount of the potash (last of the three figures given on fertilizer bags). A nutritional spray once or twice a year will aid in keeping your plant healthy, too.

Q. What could be hurting my purple and my cardinal-red passion flowers? I've sprayed them with Malathion and given them 6-6-6 and iron with no luck. The flowers of both these plants are less than half their normal size and many drop off before blossoming. Help!

A. Go after the leaf miner with Sevin or Diazinon, and the fungus with a fungicide like Captan. Passion vines are notoriously susceptible to nematodes, and there's a possibility that the stunting and bud drop may be due to root damage that won't permit fertilizers you've applied to enter the plants.

If you'll scoop up a pint of soil with some roots included and take it to your County Cooperative Extension Service office, the agent will be happy to diagnose it for nematodes.

Q. I have two pittosporums flanking my entryway and they're not making a good appearance. The older leaves are dappled with little square and rectangular yellow spots that gradually take over and turn some of the leaves all yellow. Then they fall off. What's the matter?

A. The little squares have a long name, *Cercospora pittospori.* The fact that you're losing leaves means the infection is severe, so pick off and burn the spotted ones and spray the plants with copper or Dithane M-45.

Q. Can a beverage resembling coffee be made from the wild coffee plant native to southern Florida?

A. Leaves from Florida's substitute, *Psychotria undata,* look like the real thing and its small red berries bear a slight resemblance to coffee beans. The seeds, roasted and ground, are sold in the West Indies as a coffee substitute or additive. Although the taste falls short of conventional coffee, it's caffein-free, according to studies made in Puerto Rico. Its young pods are used in salads.

4. *Ornamental Trees*

*For special index to this chapter,
see back of book.*

Trees are the crowning glory of the tropics — from the royal poinciana, one of the world's most spectacular, to beautiful oddities like the thorned and buttressed kapoks.

Many are flowering, and more of these outstanding specimens from around the world can be planted in Florida than anywhere else. They are seldom seriously disturbed by pests or diseases.

There are trees for waterfront homes, shade trees, strictly ornamental trees, deciduous and evergreen trees, trees for shallow soils and deep sand — and lists and publications on them are available at your local library and bookshops. The natives, too, are gaining in favor, especially among homeowners who want to plant and forget. Tree planting is a public service as well as a private pleasure, and County Extension Services as well as state foresters are happy to help you with advice on keeping your trees growing.

Q. We've read Edwin Menninger's books on flowering tropical trees and are just stunned by the huge number that will do well in Florida. We have no idea which of these are easy to find in nurseries and could use some help in narrowing down our choices.

A. It's hard to play favorites with these beauties since they are so distinctively different, but a purely personal list might rank the red-orange royal poinciana *(Delonix regia)* first, followed closely by the purple-and-white orchid trees *(Bauhinia blakeana, B. purpurea,* etc.), the gold-flowered silver trumpet tree *(Tabebuia argentea),* the golden shower *(Cassia fistula),* pink, yellow, or white frangipani *(Plumeria* sp.), the pink trumpet tree *(Tabebuia pallida, T. pentaphyllum* hybrids), the orange

African tulip tree *(Spathodea campanulata)*, the orange (Geiger) and white (Texas wild olive) cordias *(Cordia sebestena* and *C. boissieri)*, and the red bottlebrush tree *(Callistemon viminalis)*.

All require little or no care in South Florida, and many thrive in central Florida. They are widely available in nurseries.

In central Florida, and some parts north, the lavender-blue jacaranda would have to be added to the top of the list for showy trees, but in South Florida, the combination of warmer winters and high-pH soils makes it less spectacular.

Garden club plant sales, Fairchild Tropical Garden distribution sales, and the *Florida Market Bulletin,* distributed free by the State Department of Agriculture in Tallahassee, often offer unusual trees.

Clusia rosea — the Scotch-attorney tree — is a well-behaved choice for almost any light situation around a South Florida home. Left unpruned, it produces fascinating ropelike roots from the lower trunk. Tempermental in flowering, it has one of the most striking of blossoms — green-centered, waxy white, with a halo of pink across the center of the petals. Here it is shown espaliered.

Q. I'd like your recommendations for a few good shade trees — not too big — that have some "character" to them. In other words, I don't want just leaves!

A. Scotch attorney *(Clusia rosea)* is a well formed shade tree. The bischoffia and mahogany are widely available and give dense shade, but they can become quite large. (The bischoffia may cause root problems.)

A tree that combines shade and flowers is *Peltophorum inerme* with its showy yellow blooms from May to August. *Thespesia populnea* is a fast-grower with hibiscus-like flowers that start the day yellow, then turn maroon.

The pongam is another nice choice, along with the tamarind. You can add your own interest to a live oak by "planting" it with orchids or bromeliads or both, and its filtered light permits a variety of plantings around it. Don't overlook fruit trees as possibilities, either. Grapefruit or tangelo might be used, and the avocado is a possibility, although it attains a good size. The sapodilla has beautiful foliage and fruit which tastes like cinnamon-flavored pears.

Black olives are popular when small but as they get larger, they lose the tiered look that gives them their early character and they become "just one more tree."

Any tree will seem to be slow growing when you are waiting for shade. You can get them in sizes ranging from very small to almost mature. If you want shade quickly, you can consider buying as large a tree as you can afford. Or you may want to plant a smaller one and watch it develop.

Q. When is the best time of year to plant a tree here? Up North we set them in the ground in the dormant season, but in Florida I understand it's just the opposite.

A. Ideally, the beginning of May and the rainy season presents the most favorable conditions, but you can plant a tree anytime — provided you can take care of it. During the dry season, more attention has to be paid to watering. It's a good idea to brace your tree with supports, too. In the winds of February and March, a newcomer trying to establish itself can be easily toppled.

Farther up the state, woody ornamentals less tropical in nature than South Florida's trees can be transplanted with less attention. But ignore a tree here in the so-called dormant season (there's still root activity despite slowed-down leaf production) and it's likely to become a hatrack in short order.

Q. There's a spot in my yard where I'd like to plant a tree that doesn't need pruning. Is there such a tree in Florida?

A. Gumbo limbo is an exotic choice. Its foliage doesn't get out of hand and it has one of the most beautiful trunks in the tree world — a bright reddish-tan color. No appropriately planted tree needs pruning, however.

The thin, smooth bark of the branches look freshly shellacked, and the gnarl in the limbs gives it added interest. The gumbo limbo grows exclusively in Florida from Cape Kennedy and Tampa Bay south.

Q. I set out a "brand-new" bottlebrush, followed standard procedures, and still the tree lost several branches. Where could I have gone wrong?

A. Newly set out callistemons often have portions that die back for "no reason." Until a control is developed, horticulturists advise removing dead branches and treating the tree with a pruning paint containing a fungicide.

Q. What's the oak called that we see here in Florida?

A. It's called White, Chapman, Durand, Southern red (Spanish), blue-jack (upland willow), turkey (scrub), laurel, overcup, blackjack, swamp chestnut (basket, cow), chinkapin, myrtle, water, Shumard, post, sand post, live, and sand live.

Of these 18 different kinds, the Chapman oak spreads its shrubby growth as far south as the Everglades, the turkey oak ranges across dry pineland and sandy ridges from Collier County north, and the myrtle oak grows on sand dunes and scrub from Dade northward.

The sand live oak can be found in dry soil as far south as the 'glades and the live oak thrives throughout Florida. The rest of the trees are "Northerners" in Florida.

Q. I'm hoping you can tell me where I can find two trees that repel insects such as mosquitos and fleas. They are *Gliricidia sepium* (called madre-de-cacao) and the cajeput. I can't step outside in the summer without getting bitten.

A. *Gliricidia sepium* grows throughout the Caribbean area, and its seeds are supposed to be a rat poison. The cajeput (also known as the melaleuca and punk tree) is an introduction from Australia that has propagated itself so readily in South Florida it is choking out native growth in the wilds.

Claims for either as mosquito repellents are not valid. Clouds of mosquitos have been seen in, around, and under melaleucas.

Q. Why should only half a tree appear half dead, the other half, healthy? About three years ago this oak tree in my backyard began losing its leaves on one side and finally most of the branches seemed dead.

It tries to make a comeback but new leaves appear only to wither and die. Is it possible that lightning or weed killer could be at fault? Half the tree is doing fine.

A. Most cases of this type can be traced to construction work on driveways, walls, or foundations as far away as 20 feet. Disruption of the tree's root system on one side can permanently damage a tree.

Other possibilities are a magnesium deficiency, oak galls or burls (branches with these should be pruned and destroyed), or scale, but these would be less likely to restrict themselves to half your tree. The two possibilities you speculated on could be at fault, too.

Q. We are having trouble with the large old sea grape tree in our yard. Very few of the leaves come out as large as the first ones on the enclosed sample, but nearly all are small with red squiggly lines on them. What can we do to doctor it?

A. This damage is caused by a weevil that has previously only established itself on pigeon plums, close relatives of the sea grape. Spray the tree with 50 percent Sevin wettable powder at the rate of two tablespoons per gallon of water, repeating the application every ten days for three applications. The ground around the tree should also be sprayed three or four times.

Q. Every night, from nowhere, millions of beetles invade my mahogany tree. When the swarm leaves, the foliage is riddled with holes and the tree has a scrubby appearance. What are these beetles and what can I do to control them?

A. These night flyers are Cuban May beetles. At daybreak they burrow back into the ground. Their other victims include roses, avocados, and mangos. Seasonal pests, they can be controlled by spraying your tree with Malathion. You may want to hire a professional sprayman who can use a stronger material but be sure he is familiar with the problem.

Q. What's the matter with my bischofia? It's losing its leaves but I can't see any bug problem on it.

A. For several trees, spring is fall in Florida. It is normal for the bishofia and natives like the mahogany and pongam to shed at this time of the year.

Q. Something's fouling up the leaves on my bischoffia tree. They turn red and drop. Seems like there are some little circular bugs on the undersides of the leaves that are plentiful enough to be suspect. What do I do?

A. Oleander scale fits your description and may be controlled by spraying with Cygon or Malathion.

Q. I've just nursed my Norfolk Island pine back from a fungus or blight that caused the lower branches to turn brown. Now the bark is splitting and I'm about ready to chop the tree down for firewood. What should I do?

A. Here's one of the rare instances where a tree that has been fertilized well and is in good, strong growing condition is worse off then its undernourished sister. Such vigorous plants are at a disadvantage in coping with the physiological stresses caused by uneven weather conditions. Bark splitting and peeling are often brought on by periods of wet weather following periods in which rainfall has been below normal.

Best thing to do is fertilize lightly and frequently to steady the growth rate and return the tree to even development.

Q. I'm having a problem with my Norfolk Island pine that only affects the small new branchlets. They turn yellow and drop, leaving a bare twig. But the rest of the tree looks all right. What do I do?

A. It may be that you have been overwatering your tree. This sort of damage may appear weeks after excessive watering. If the twigs are not severely damaged, they will produce new sprouts as the season progresses. Frequent, light fertilizing should help remedy the situation.

Q. My pandanus tree doesn't look well even though I fertilize and water regularly. I'm hoping from the enclosed leaf sample you can tell me what's wrong.

A. Your leaf doesn't look bad — no evidence of bugs or bug damage. When diet deficiencies, insects, and root damage have been ruled out, general decline of a tree can sometimes be attributed to a fungus or blight. Try spraying your tree with neutral copper three or four times at two-week intervals.

Q. Why doesn't my black olive bear good-tasting olives? I feed and water it regularly.

A. It is not a true olive. *Bucida buceras* is not even a relative of the edible olive, *Olea europaea,* which grows only in areas with its native Mediterranean climate. The bucida fruit only looks more or less like the real thing.

Q. My Italian cypress is on a rapid decline. Its leaves are dying back and have a rusty appearance. I really can't see what's causing the problem. Does it sound familiar?

A. Italian cypresses are susceptible to attacks by mites and fungus — sometimes a combination of the two. To be on the safe side, apply both a miticide and a fungicide — Kelthane and Dithane or Maneb. The sooner you are able to spot mites, the better for easy control. Next time you can't seem to locate the source of damage, take out your magnifying glass and look for the spider-like bugs, especially on the underside of leaves.

Q. I have an ailing schefflera covered with spotted brown leaves. Each spot has a yellow "halo" but they're giving my trees a devil of a time! Eventually, the leaves turn brown and fall off. What's the cure?

A. Sounds like the infection is alternaria leaf spot. Crowding of plants, high humidity, and overhead watering can cause a rapid buildup of the disease, especially on younger plants. To get rid of the fungus, spray with Zineb or Manzate-D on a weekly schedule for about a month. Separating plants to allow better air circulation will discourage it.

Q. My schefflera has several leaves afflicted with a strange corklike growth on the undersides. On some of them, almost the entire back of the leaf is covered. What is this and how can I get rid of it?

A. It's called edema and also shows up on the older leaves of aralia, ligustrum, and other plants. The raised corky tissues are often grouped together and appear following periods of heavy rainfall or too-frequent waterings.

Edema is most common in the late fall and early winter. No control is necessary except to follow a moderate watering schedule and fertilize regularly.

Q. My umbrella tree has been losing many of its leaves, which appear to have a scale. What must I do to correct this condition?

A. Your schefflera probably has a pyriforme scale. Spray with Malathion.

Q. Something fatal is happening to my schefflera. It looks limp and has black spots. What do I do now?

A. Sounds like the fungus disease, alternaria leaf spot. This problem most frequently is found on plants already in trouble — scheffleras in pots with inadequate drainage, plants that have been over- or under-watered or damaged by too much fertilizer or insecticide burn. Using a liquid fertilizer such as a 20-20-20 could give the plant a boost, too.

Q. My traveler's-palms were successfully grown from seed and did well until they reached about 30 inches. After that they slowly died. I put on Epsom salts and 6-6-6 without success. Finally a generous amount of fish was buried deep around the root area. This has barely kept one alive. Please advise.

A. If you're on salt water marl fill you may be having both aeration problems and nutritional deficiencies. Mixing sand into your soil will improve drainage. Nutritional sprays may benefit, too.

Traveler's-trees (they're not palms but cousins to the banana and strelitzia) are one of South Florida's most dramatic landscaping materials. They like full sun and soil high in organics. They grow best located away from salt air and should be fed several times during the growing season.

Cercospora leaf spot is their main nemesis. It can be routed by spraying with Captan, Ferbam, or copper compounds.

Q. Our neighbor has a swimming pool and has asked us to remove our rather large ficus tree fearing the roots might damage the pool. I've been told it might also damage my house, the canal wall, septic tank, etc. Should we remove it?

A. Roots from this tree will go quickly and powerfully to sources of moisture and nutrients. They also play hob with sewer lines and septic tanks. The possibility exists that the problem could become severe enough to weaken a foundation to the point that winds of a hurricane might finish the job. Lovely and popular as this tree is, it causes considerable damage by its strong root system.

There are many lovely trees that flourish in South Florida that do not present this problem. Citrus trees are usually safe around pools since they develop a tap root (which usually goes straight down) in addition to many surface roots.

Q. I have a ficus hedge, and about 85 percent of the leaves have been severely damaged. I can see no worms, bugs, or anything that would cause this condition. Would you please enlighten me as to what causes it and what I can do to remedy this condition?

A. From the chewed leaves what caused the damage isn't obvious, but in recent years there have been increasing reports of this problem caused by the Edwards moth caterpillar.

This pest begins feeding on the ficus when the worm is small. After eating its fill and growing larger, it leaves the tree or hedge to build its cocoon on a nearby structure.

For control, the small worms would have to be discovered before they could cause much damage. Spraying with Sevin should stop them. Follow all directions and warnings on the label.

Q. The former owners of our property must have loved dense shade. The foliage on the loquat trees they planted is so thick that we find them depressing, even though we love them as young trees. My brother says cut them back severely in a rounded shape, I say this will simply make them grow out denser than ever. Right?

A. Right. But there is no reason why they cannot be opened up to the light. Careful pruning will make them even more effective and beautiful than young trees. The thinning must be done carefully so that the balance is maintained and the structure of the trees is not weakened. Cuts must be made closely and treated so they will callous and heal over. If you have a good eye for line you could do this. Otherwise, it would pay you to let a tree surgeon take over the operation.

Q. I have a frangipani that has five limbs but only blooms on one of them. For three years I tried to force flowers on the others with liquid fertilizer but I only got more flowers on the same branch. Why is this?

A. If the flowerless branches are also leafless, they may just be dead, but otherwise they could be in a vegetative rut. Occasionally branches fail to mature into reproductive growth. Sometimes this is further restricted to only branch tips. Known as differential response, it can under certain conditions be attributed to the plant's root system.

Q. How do you grow a cecropia tree? I like to use the dried leaves in arrangements and am hoping I can raise my own.

A. Cecropias need a sheltered location, faithful watering, and an annual fertilizing if planted in "good" soil. The more alkaline the soil, the more frequently they should be fertilized. Broken shade is best. These open, tall trees are virtually pest-free; ants may be found in the hollow branches of old trees but apparently do no damage.

Q. I'd like to know what's the matter with the enclosed bottlebrush leaf. Do the spots cause the yellowing or the yellowing, the spots?

A. Unless the brown spots cover many leaves, try strengthening your tree by clearing up the nutritional deficiency that the yellowing indicates. This is called chlorosis and shows that one or more of the so-called minor elements — manganese, zinc and iron — are missing from the tree's diet. These can be applied as foliar sprays or in chelated form to the ground.

Fertilizing monthly with 6-6-6 until the plant has more vigor may make applying a fungicide unnecessary. When the leaves are greener, the tree will need fertilizing only four times a year at three-month intervals.

Q. I'd like to plant some flowering trees around our new home, but since we're in a pretty stark-looking development, I'd really prefer fast growers. Is there such a thing among flowering trees? My neighbor says most of them are slow.

A. The general interpretation of a fast-growing tree is one that will reach 12 feet within three years. There are a number of flowering trees that are slow starters the first two or three years, then shoot up five or ten feet, sometimes more, in a year. The shaving brush tree is one, so is *Bombax ceiba* and other members of the bombax family.

Faster are the golden shower tree *(Cassia fistula)* and gold-flowered silver-trumpet *(Tabebuia argentea)*. Speedier still are the members of the callistemon family. The red, weeping bottlebrushes hold their own among the world's best ornamentals, and the melaleuca — a yellow-white bottle-brush relative — is not only fast growing but almost free from disease and pests. Some people, however, are allergic to its flowers.

Q. Why hasn't my jacaranda tree ever bloomed? It is eight years old and we've given it excellent care.

A. Chances are good that your tree is a seedling and, if this is the case, it might take as many as 20 years to bloom — if ever. Try to find a grafted jacaranda to make sure you get blooms sooner.

Q. I want to build a brick patio and shade it lightly with a poinciana. How close can I come with bricks to the base of the tree without running into root problems later?

A. A four-foot circle of earth around the base should be enough room to allow for tree growth and aeration of a part of the root system, an important factor to be considered when laying surfaces around patio trees. Ordinarily, the roots of a poinciana will not raise brickwork beyond this area.

The world's most flamboyant tree, the royal poinciana, thrives in South Florida with no care. It varies in color from light orange to rich red. To preserve the natural umbrella canopy of the tree, major branches should not be pruned.

Q. Is there any preventive for the small worms that infest my poinciana tree? Bark from the trunk peels revealing worm nests. The worms consume the leaves. What to do?

A. All the loose bark should be scraped away to the good wood. Then it should be sprayed with a fungicide and sealed with a pruning paint. Then the entire tree should be carefully sprayed with Sevin.

Q. Could you give me some information about the bottlebrush tree? I'm interested in when to plant, rate of growth, and any special requirements.

A. A container-grown plant can be set out at any time in South Florida. Just be sure that you can and do supply plenty of water to help the tree become established.

Plant growth varies considerably depending on the care given. With a good program of fertilization and irrigation this tree should grow rather quickly. A light application of garden fertilizer monthly during the first year it is in the ground should get it off to a good start.

The tree has no unusual requirements, but should have full sun for best growth and flowering. The bottlebrush (callistemon) is usually too weeping to be in the shade tree category, but is good in a shrubbery border or as a specimen tree.

Q. Have you ever heard of people eating the seed pods of golden shower trees? I have a friend who collects them and claims they have medicinal virtues.

A. Some say that their fiberous interior has a catharic value, others just like the unusual taste. The part they munch on is the padding between the seeds inside.

Q. How can I give a small tulip tree that extra push it needs to get started?

A. This tree *(Liriodendron tulipifera)* must have moist soil for good growth, so watering should not be neglected. Monthly applications of a fertilizer with a high nitrogen content or organic matter will help. A layer of mulch also is good; however, these trees do far better farther north.

Q. The past two summers my favorite tree, a frangipani, has started shedding its leaves in July. The problem seems to be a red-colored blight under the leaves. I have tried two different sprays — one with an oil base — and neither made any difference. Now the tree's getting moldy. Is there a solution to both ailments?

A. Frangipani rust will spot leaves brown and its control is a neutral copper solution applied monthly throughout the summer. The lichens cause the "mold" and they are usually on older trees. Not harmful, they can be kept to a minimum, though, with regular fertilizing and watering. Most gardeners don't bother controlling the rust as it worsens at a time when most frangipani are ready to lose their leaves, anyway.

Q. I have had a lovely dogwood tree that I raised from seed I brought with me from Alabama. This fall it has taken a turn for the worse. The leaves are turning brown and the whole appearance is mangy. What's the matter?

A. The problem is South Florida and a fungus, plus marginal breakdown that has been plaguing other dogwoods in Dade County for several years. A fungicide might help but it's no insurance that the tree will make normal, vigorous growth in this area. As you undoubtedly know, trees used to annual freezing temperatures do not grow well — if at all — in South Florida and you're taking a big gamble when you lavish attention on them.

Q. How many kinds of tree hibiscus are there? I thought there was only one but my neighbor's has larger leaves and flowers.

A. *Hibiscus tiliaceus,* the seaside mahoe, has yellow flowers in the morning, pink by evening, and velvety fruit with persistent calyces (those modified flower leaves that hang on). *Hibiscus elatus* has larger leaves, flowers that turn from red to dark maroon, and fruits without calyces. *Thespesia populnea,* or the portia-tree, has leaves up to five inches long and flowers that change from yellow to purple in a day.

Q. My husband and I are having an argument over orchid trees. I say the ones most common in Miami are the Hong Kong type. He insists the trees in bloom along Old Cutler Road are *Bauhinia variegata*, but he's forgotten why. Our books say they're both purple.

A. *Bauhinia variegata,* a small Indian tree also known as mountain ebony and the "poor man's orchid," gets most of the credit for decorating streets from Vero Beach south. Its two-tone or three-tone purple, red, and white flowers give it away. It is a spring bloomer.

Bauhinia blakeana or the Hong Kong orchid tree is a bigger, more cold-sensitive, more fragrant tree with richer, reddish-purple flowers, five and half to six inches wide, sometimes verging on crimson. Its blooming season is October to March. It was propagated originally in Florida from budwood brought to the University of Florida's experimental station in Homestead and is considered the showpiece of the orchid tree family (no relation to the orchid plant).

The white orchid tree is a variety of *B. variegata* called 'Candlda.' *B. galpinii,* the brick-red orchid tree, blooms from spring to fall. These trees are not related to the corsage-type orchid.

Q. The growing tip on my ponytail palm is dead, but there are still new shoots growing from the sides. Can these shoots be removed and planted, and if so, how?

A. *Beaucarnea recurvata* (not really a palm) grows best from seed, but if you think the tip is dead you have nothing to lose by cutting the shoots off closely and planting them in peat and sand mixed together. If you can rig a plastic bag cover supported by a couple of sticks, they'll have a better chance.

Anyone removing shoots from a healthy ponytail is taking a chance and definitely should treat the wound with a pruning paint.

Q. I would like to have some more information on the Florida maple. Is this a maple that doesn't lose its leaves in the fall?

A. No, it's a card-carrying, deciduous maple, *Acer rubrum.* One of several kinds of maples native to Florida, it may be found on moist or wet ground in every county. The red and yellow foliage appears consistently in the fall, drops, and then is followed by red flowers and fruit clusters in the very early spring.

Don't expect your nurseryman to carry this tree. Occasionally it can be purchased at garden club plant sales. It doesn't need a swamp to grow in, many are planted around the area on high ground.

5. *Palms*

For special index to this chapter, see back of book.

Among the thousands of varieties of palms there are species varying in height from a few feet to 100 — many represented in the world famous palm collection at Fairchild Tropical Garden in Miami.

The most popular in South Florida include the tall coconut (now being replaced by the Malayan, resistant to lethal-yellowing disease), the royal, adonidia, areca, Alexandra, cabbage, paurotis, pigmy date, MacArthur, and chamaedorea palms. It has been found that many of these palms are susceptible to lethal yellowing, and anyone contemplating the purchase of a palm would be well advised to contact a county agent to determine if the tree is on the list of those that could get the blight.

The county agent's office also is a source of free information on identifying the blight and injection directions for using the drug oxytetracyline either as a preventative or for remission of symptoms.

Although seldom noticeably bothered by problems, palms usually are not treated until pests or nutritional deficiencies become severe. Look up in the air often so that you'll be able to come to their aid in time to preserve their beauty.

Q. What are the most popular palms here? I'd like to start a collection.

A. George Stevenson in his book *Palms of South Florida* gives the 10 most widely planted — but they were popular before the arrival of the incurable (although treatable) lethal yellowing blight.

Here are the most often seen:

Coconut palm *(Cocos nucifera)*, cabbage palm *(Sabal palmetto)*, Manila, adonidia, or Christmas palm *(Veitchia merrillii)*, royal palm *(Roystonea elata)*, areca palm *(Chrysalidocarpus lutescens)*, pygmy date palm *(Phoenix roebelinii)*, fishtail palms *(Caryota* sp.*)*, Canary Island date palm *(Phoenix canariensis)*, queen palm *(Arecastrum romanzoffianum,* formerly *Cocos plumosa)*, and Washingtonia palms *(Washingtonia robusta)*.

Q. What kinds of palms can you grow in containers?

A. Chamaedoreas — the bamboo palms — probably are the most widely

Residents of Coconut Grove gather to hear University of Florida lethal yellowing researcher Dr. Randy McCoy explain effectiveness of oxytetracycline and demonstrate palm injection. Palm owners should contact their county agents for detailed help.

grown potted palms, and they do well in light soil. Licualas and pinangas like pots, and the rhapis palms are frequently seen in containers indoors.

Others suitable include arecas, cocothrinaxs, howeias, *Phoenix roebelenii*, reinhardtia, *Chamaerops humilis,* and *Drymophloeus beguinii*. Easiest of all these to locate are the chamaedoreas, cocothrinax, rhapis, and phoenix palms.

Q. What are the symptoms of lethal yellowing, and how many doses of blight drug should you use for each stage?

A. The first visible symptom on coconuts is the dropping of immature nuts, followed by blackening of the new flower stalks. Then the lower fronds start turning a brilliant, solid chrome yellow. Yellowing, then browning, spreads up the tree to the center spear leaf, then the bud turns brown and the crown falls off, leaving the "telephone pole" stage.

Trees with more than six leaves yellowed or browned from lethal yellowing seldom recover after treatment. Six leaves require six doses (packs) of Uri-Tet or six slightly rounded teaspoons of Terramycin. The doses required are graded down proportionately until only one dose is needed as a preventative on a tree showing no symptoms. Preventative treatments have the highest rate of success. Many people prefer to give a tree showing any symptoms the full six doses if money is not a crucial factor. It is recommended that all palms — even those not on the list of known victims — be given a preventative dose once every four months.

After treatment, the tree may continue to decline since it can take some time before improvement shows. It should not be cut down until the second treatment four months later produces no improvement, and all fronds are yellow. In many of those cases where owners say treatment was not effective, it almost always is found that there was some slipup in following injection instructions to the letter.

Q. What palms are susceptible to lethal yellowing?

A. In addition to the coconut palm *(Cocos nucifera)* and Christmas or adonidia palm *(Veitchia merrillii)* — the two most widely planted palms in South Florida — the following palms have been reported with symptoms of lethal yellowing:

Fiji Island fan palm *(Pritchardia pacifica),* Thurston palm *(Pritchardia thurstonii),* Arikury palm *(Arikuryroba schizophylla),* Talipot palm *(Corypha elata),* Canary Island date palm *(Phoenix canariensis),* commercial date palm *(Phoenix dactylifera),* Chinese windmill palm *(Trachycarpus fortunei),* fishtail palm *(Caryota mitis)*, Palmyra palm *(Borassus flabellifera)*, spindle palm *(Mascarena vershaffeltii)*, and *Phoenix reclinata*.

This is not a complete list as additional susceptible species continue to be identified.

Q. My brother is a doctor and I can obtain Terramycin from him at a discount. How should this be mixed for use on palm blight?

A. It shouldn't. Terramycin available from veterinarians or physicians is a special compound made to mix with blood. The tree formula, entirely different, is made to move through plant cells.

Q. What's the best way to treat a tree that has lethal yellowing — and the cheapest?

A. It's recommended that anyone who wants to treat trees without professional help contact the local county agent's office and obtain its free literature on the problem, first.

Two devices are currently available for do-it-yourself treatment — the plastic Mauget capsule and the Minute Tree Injector gun, which looks like a large hypodermic needle.

The Mauget is good for one dose only — a dose being five grams of Terramycin or Uri-Tet, both equally effective. Each does contains one gram of active ingredient (oxytetracycline) and four grams of buffers to keep the material from burning the tree and to introduce it effectively.

The Mauget must be thrown away after one use. It may sell anywhere from $1.50 empty to $1.50 to $3.00 filled. It consists of a plastic container about two inches tall and an inch-plus in diameter, and a hollow metal tube.

The container is in two parts. One half is filled with one pack or one rounded teaspoon of powder and water. This half is held upright and the empty half is fitted over it. It is shaken for a minute, then you step on it so the top covers all but about a quarter-inch of the bottom. This builds the 30 pounds of pressure necessary to force the liquid into the tree.

Then a hole is drilled into the tree and the metal tube is pressed into the hole. It should not be hammered into the hole as the hammer will distort the soft metal, which will cause the injector to leak. It can be very slightly soaped, then pushed in with a wood block. The hole must not be too large (specific instructions for drill bit size are given with the injectors) or the metal injector made too slippery or the injector will not sit firmly enough in the hole.

Next, the metal injector is placed into a recessed, sealed hole in the base of the plastic capsule and pushed hard so that the injector punctures this seal.

The capsule is left on the tree for a day — longer for the Christmas palms which have denser wood — and then removed. Any liquid remaining can be dripped into the hole with an eyedropper. The hole in the tree should be plugged with a twig. If there is repeated leaking from the area around the injector where it enters the tree, putty may be used successfully.

The disadvantage with the Mauget is that six holes and six capsules must be used for a tree in the most acutely treatable stage. It is an easy procedure — but usually only after a few practice capsules have been wasted. It is the cheapest method if the Maugets are purchased where they are available filled for $1.50, and if there are only a couple of trees needing one preventative dose or one tree needing not more than three doses.

In the long run, however, and if you have several trees, the Minute Tree Injector (retailing for about $15) is much simpler. Drilling the hole is still required, but the material can be hand-pumped in within a minute. It lasts well over 70 shots — and some have reported as many as 200 shots from it.

Anyone who is timid about trying either method should shop for an inexpensive blight treatment service and allow them to treat one tree, observing the procedure.

An important note: if there is any leaking, the procedure should be repeated as correct, complete dosage is vital to success.

Q. What are the symptoms of lethal yellowing on palms other than coconuts and is the method of treatment the same?

A. The symptoms vary, and with some there is not the yellowing characteristic of the mycoplasma-caused symptoms on coconuts. (Mycoplasmas are a cross between bacterias and virus.) Contact your county agent for specific information.

Ask your agent also for the best way to treat other palms. With multiple-trunked palms, for instance, shooting three of the trunks may be all that is necessary.

Q. I'd like some information about the Malayan coconut. Can it get the blight, is it a dwarf form of the common coconut, and what do the colors refer to when it is sold as green, yellow, and red?

A. The gorgeous Malayan coconut palm is 98 percent resistant to lethal yellowing — and that's for all colors. This figure is confusing to many. What it indicates is that the huge majority of nuts are 100 percent resistant but a certain small number, like the result of any hybridization, may be genetic throwbacks to less resistant strains.

The Malayans propagated by all reputable nurseries are from Jamaica;

they are certified resistant by the Coconut Industry Board, which has thought enough of their tolerance to use them as the stock for replanting Jamaica's blighted copra plantations.

Nuts purchased from other islands or gathered from old trees in Florida are not guaranteed resistant. Malayans cross-pollinate easily in nature with the blight-prone coconut, and the resulting nuts will produce trees not as resistant.

Malayans are not dwarfs, although that designation is still found attached to them on occasion. Their nuts tend to be smaller and their trunks somewhat more slender, but they'll grow as tall — as witnessed by the tall specimens on the west side of Parrot Jungle and the east side of the U.S. Department of Agriculture's horticulture research unit on Old Cutler Road in Miami.

Malayans do like to be watered, and those that aren't treated like a weed will grow taller faster. The colors refer not to the leaves but to the leaf stems and the nuts, which are a pure and brilliant green, yellow, or gold, sometimes called "red."

Q. I'd like to take my chain saw and smooth out the ruptured-looking bases of two of my coconut palms. Would this hurt the trees any? They appear in good health, except for this unattractive condition, which seems to be pretty common.

A. This odd proliferation of the coconut's root system is often found on shallow limestone soils; the roots are literally growing up out of the ground. Palms, like most trees, are sensitive to mechanical injury to their trunks and you might regret interfering with this compensatory move on the part of your trees.

Q. Can you tell me if it is possible to root the side shoots growing out of the base of a fishtail palm? Recently I was given a specimen about six feet tall with five or six suckers, and I would like to know if these can be removed and propagated.

A. This palm often is propagated by dividing the plant, but in doing so the original palm may be sacrificed. However, if you can remove a sucker with some roots to sustain it, you may be able to start new plants without damage to the original.

Q. We just returned to Ohio with souvenirs of our two-week Florida vacation, and now we want to know how you sprout a coconut.

A. In the tropics it can't be raised indoors like a philodendron. It needs warm, humid air outdoors, preferring, in the U.S., the coastal areas of South Florida. (It will grow in interior South Florida if given maximum protection.) Sprouting procedure is to plant the coconut (leave it in the husk) on its side, covering it not more than two-thirds with soil. Completely covered coconuts will not sprout. The soil should be kept moderately moist and the young tree shaded from the sun in midsummer and protected from frost in the winter for maximum success.

A quarantine has been placed on coconut palms by the Florida Department of Agriculture, prohibiting the removal of plants or nuts from areas infested with lethal yellowing, so — technically speaking — you have an illegal nut.

Q. Is there something you can do to make coconut palms bear coconuts more quickly? We have four trees we raised "from nut" five years ago and they haven't produced any coconuts yet, although they did bloom for the first time last year. And, ordinarily, how long does it take from blossom to mature nuts?

A. Fruiting depends on several factors, foremost of which is the type of coconut. Fruiting of the large varieties usually begins in seven to eight years from planting; the hybrid Malayan palm starts bearing in about five years.

The flower clusters usually appear the fourth or fifth year, but some trees may not bloom until the seventh or eighth year. It takes about 14 months from blossom to mature fruit.

Results of research recently conducted by the Bureau of Soils, the United Nations Special Fund Project, and the Philippine Coconut Research Institute indicate that coconuts receiving fertilizer consisting mainly of potash grew faster than those not receiving the potash boost. They also produced fruit earlier and in three times as much quantity.

Potash also affected the size of the nuts. Without potash, eight nuts are needed to make a kilogram of copra. With potash, less than four and a half nuts are needed.

Test trees planted three years ago and given a potash-high fertilizer grew very rapidly. Older plantings showed that potash-fertilized trees started bearing in four to five years — three to five years faster than the average tree.

A 1-1-1 ratio fertilizer such as 6-6-6 or 8-8-8 is adequate to raise a healthy coconut but if you want a super-palm, feed it with a 1-2-2 analysis fertilizer such as 4-8-8 for more and bigger nuts.

Q. My coconut palms look horrible. The rest of my garden is busy putting out new growth but the palms are really dragging their heels. They've been going downhill since winter but they don't have lethal yellowing. Is there any hope for their recovery?

A. Their condition could very well be another side effect of a winter chill.

A long period of sustained cold can hold back many tender palm trees and ornamentals.

Most subtropical plants don't grow appreciably when temperatures drop below 55 degrees. As a result, they experience a slowdown and drop many lower leaves. The root system becomes less active in picking up food and water from the soil to replenish moisture and nutrients spent in normal growth and maintenance.

Another deterrent to vigorous spring growth can be several periods of heavy rainfall which wash away much of the food stored in the soil and cause even mature plants to become very deficient.

Special care should be given to these plants to stimulate them into spring action with extra amounts of water and fertilizers. Gardeners should also be on the alert for a bud rot fungus disease which sometimes develops on coconut and adonidia (Christmas) palms following periods of long exposure to cold weather. Forcing fast spring growth will discourage its development. But as a protective measure or if symptoms of bud rot appear, you should spray with a neutral copper fungicide.

Q. How can you tell bud rot in coconut palms from lethal yellowing?

A. In lethal yellowing, the tree declines rapidly, the lower leaves turning a bright yellow, and the color progressing upward to the heart leaf, last to go.

With phytophthora bud rot, first reported on Grand Cayman Island in 1834 and in Florida in 1924, early symptoms of infection are found on the young developing fronds. Brown sunken spots, yellowing with or without spotting, and withering of the fronds indicate infection. The fronds turn a light grayish brown which becomes darker brown as they bend over and collapse at the base. Infection spreads inward to the soft tender bud tissue and outward to the base of the surrounding fronds, which eventually turn yellow and fall to the ground. Trees which are bearing may drop the young nuts but retain the older nuts until maturity.

The most obvious symptom of bud rot is the dying and subsequent loss of the bud in the late stages of the disease. Also, a foul odor is usually associated with the decaying bud tissue. Periods of high humidity favor the development of this disease. Bud rot is most commonly found one to two months after periods of heavy rain. Infection occurs on trees two years old or older. The trees may die from two months to two years after infection, but natural recovery has been reported occasionally.

Spores of the fungus are broadcast by wind, wind-borne rain, and possibly by insects. Infected trees left standing may help spread the problem by harboring spores.

As far as a control goes for bud rot, the common neutral copper fungicide plus a spreader-sticker has been reported to help when applied to adjacent healthy palms or to palms showing early symptoms of the disease. Removal of fronds showing early symptoms of infection is also recommended. Palms showing advanced symptoms should be destroyed.

Q. My coconut palm is looking droopier than ever and I need help fast. Some of the fronds are turning yellow, then falling off. The tree gets good care but I don't overdo. I don't suspect lethal yellowing because the flowering stalks look good and young nuts aren't dropping. Any ideas on what's the problem?

A. Could be palm scale. If you notice ladybugs on the fronds, they may be present in numbers ample enough to do the job of cleaning up the scale for you. Otherwise, spray with oil emulsion.

Q. Why are the trunks of some palms so uneven in size? When we bought our house, the royal palms in front had nice, straight trunks. We have watered and fed them regularly since then but the tops of the trunks are now swollen-looking compared to the rest of the tree.

A. Unlike woody trees, a palm that's given more attention than it was accustomed to receiving will not "put on weight" up and down its entire length. The part of the trunk that lived through the neglected period will never show an increase in its diameter while the trunk formed above this point will expand to a much greater thickness when vigorous growth is encouraged.

Q. I've got some young palm trees with problems. What do I do for large brown splotches on the leaves?

A. Sounds like the palm leaf skeletonizer, the larva of a small moth that attacks both upper and lower surfaces of palm leaves. Infested palms are seldom killed, although the dead leaf area and webbed galleries are not assets to the trees.

Leaves and surrounding husks may be removed and burned or plants may be sprayed with Sevin 50 percent wettable powder at the rate of two tablespoons per gallon of water. Thorough coverage of the infested area is necessary, and removing and burning or spraying must be continued until the insects are under control. Dusts are not satisfactory because of the problem in obtaining adequate coverage.

Q. Something's on my phoenix palm that sticks the fronds together and gives the tree a messy appearance. What is it and what can I do about it?

A. The palm leaf skeletonizer is probably the most commonly encountered pest on palms in Florida. The young caterpillars feed happily in webbed runways, often folding fronds together and munching under the protected enclosures. Susceptible palms include the saw palmetto, sabal, phoenix, Canary Island, and coconut. For treatment, see preceding answer.

Q. We have recently moved to Florida and are building a home here. We have been looking forward to planting some palm trees on our property; however, just today we were told by a long-time Florida resident that palms are a nesting place for these big black water bugs that we see so often. Seems that these bugs are under the bank of the trees near the top.

If this is true, it will certainly discourage us from planting palms. We simply detest the big water bugs and don't want to do anything that will contribute to their existence.

A. A palm tree is no place for bug haters but since you plan to live in a house and very likely will never even see the top of a palm, you're safe. Or, at least you are "safe" if you don't sleep with your doors open and you keep your doors closed and your nonrefrigerated food in airtight containers.

Many newcomers to Florida have been frightened by bug stories, and, of course, there are more bugs in the tropics. But they know their place. Undoubtedly, there are bug-prone people — some with less-than-tidy homes — whose lives have been haunted by real and imagined run-ins with all kinds of creatures, but they are in the minority.

Different kinds of bugs, including "water bugs," do make their homes in palms, but they'll live elsewhere if the palms aren't present. At least with a palm, you know where they are. (Actually, when you come right down to it, your average suburban home-grown bug is much neater than any domestic pet.)

Q. Do you know of a nurseryman who sells cabbage palms in containers small enough to move by car? About six years ago I planted some seeds of these trees. Two came up at my place on Big Pine Key and two more at the house of a friend in Fort Lauderdale. But the only thing that we have to show for these six years is large "palmetto bushes."

Is this possible — do some of the seeds turn into bushes?

A. A tree is a tree is a tree. It's just that the cabbage palm hates to grow up. At Fairchild Tropical Garden, home of one of the world's largest palm collections, Superintendent Stanley Kiem says this reluctance is so pronounced that the buds tend to grow downward instead of up for the first several years.

They'll sometimes grow right down to the bottom of a can — eight inches or more — before they turn about and start heading up. This is, however, good insurance against damage from fire and cows.

Kiem says that at his home there is a cabbage palm on the verge of developing a trunk — and it is 24 years old. The tree gets a minimum amount of care. If you were to keep yours well watered, fertilized, and weeded, it would very likely develop much faster.

Cabbage palms in a marl area near a farm, for instance, where irrigation was frequent, outstrip palms in less advantageous spots.

Still, few nurserymen are going to go to the expense of nursing along for years a palm that people want to buy as an inexpensive tree. And people haven't been educated to the value of this palm in its immature state as a beautiful ground cover — the "scrub palmetto" that covers acres of pineland.

Young palmettos are sold quickly at Fairchild's annual plant sale, however. You can order mature cabbage palms from nurseries. But these are transplanted from undeveloped 'glade land, not grown from seed.

Cabbage palms are also called sabal palms or palmettos. They look like the saw palmetto common farther upstate, only the saw palmetto has sharp toothed leaf stalks and leaves with stiffer segments.

Seeds of the cabbage palm germinate in a month's time if they are fresh (still covered with soft pulp) and weevils don't drill holes into them.

Q. I planted 20 small palm trees from cans to the ground about a month ago. I have been watering them every three days and have fertilized them once. Some say to water more often, and some say to water them very good only one time a week. Please advise.

A. For newly planted palms, your watering schedule seems to be pretty good. But you soon can start spacing the irrigations a bit farther apart.

The general rule for watering plants when they are set out is to irrigate daily for a week after planting, every other day the second week, every third day the third week and so on.

After six weeks, the plants should be fairly well established and a thorough soaking of the root zone once a week may be enough.

Of course, if you have soil which dries very quickly or we have a period of several days when it is very dry and windy, you may have to irrigate more often.

A light application of fertilizer monthly for the first year the plants are in the ground should bring them along. After that you can space out the applications.

Q. My areca palm has, apparently, the same disorder I've seen on some other palms in my neighborhood. The leaves start turning yellow, get a frizzled appearance, and parts of them go brown. The new leaves are stunted. What's wrong?

A. The problem you describe affects coconut palms, royal, fishtail, dwarf date, Canary Island date, Chinese fan, and gru-gru palms. Sometimes it's called "curly top" or "curly leaf" and sometimes it's called "frizzle leaf."

It's a manganese deficiency, and in a mild stage shows as light green leaves, the color becoming yellower the longer the supply of manganese in the soil is depleted. The leaves are characteristically streaked since the tissue between the veins is a lighter green than the vascular tissue. In severe cases much of the leaf area dies, possibly due to fungus attacking the weakened tissue. When the tree is on its last legs, it is no longer able to push out new leaves.

Manganese may be applied either to the soil or as a foliage spray. You can either broadcast it under the spread of the tree or put the manganese sulfate into several small holes made about a foot deep within the root range (but away from the trunk).

Q. Something is eating my young areca palms. We call him "Spot." He's a caterpillar with brown ends, a bright green middle, and a brown spot on his back. How do I get rid of him and, by the way, what will he turn into?

A. Don't be tempted to pet "Spot." He's a saddleback caterpillar and those nettling hairs on his sides and ends are connected to poison glands that can cause a nasty rash. Saddlebacks mature into dark-colored moths. Their hosts include roses, castor beans, and many other shrubs. Control them with Sevin.

Q. Just for fun, I'd like to try growing some palm trees from seed, and maybe some other trees people usually buy as seedlings. How long can these seeds be stored?

A. Unlike seeds from temperate-zone trees, tropical seeds do not require a period of dormancy for them to germinate. Most of them should be planted as quickly as possible after they mature. (And don't put them in the refrigerator.)

Some palms may grow after storage of two years, but they are exceptions. Lychee seeds are an example of those that should go into the ground almost immediately. If they are allowed to dry for as briefly as 24 hours they may not germinate.

Q. I always picked up some of those fat red berries from the Christmas palms. I have planted dozens, but never have raised one. Even when the seed comes up, it dries in a few days. When I dig to find the

Fairchild Tropical Garden, 10901 Old Cutler Road, Coral Gables, has one of the most outstanding collections of palms in the world. A tram tour through the park helps form preferences for home landscaping. Some of the shorter palms make good hedges, barrier plants, and indoor trees.

seed, it usually has disappeared. Does something eat them? How can I raise one?

A. It may be you are keeping the seeds too wet. If they are kept in soggy soil, they just rot.

You also may be planting the seeds too deeply. Try putting the seeds in a container of soil, just barely covering them. It won't hurt if part of the seed is exposed. Keep them just moist, but don't let them be soggy.

At Fairchild Tropical Garden, the seeds sprout readily when they drop into the layer of old leaves and other material beneath the palms.

This is one of the palms susceptible to lethal yellowing.

Q. Is it possible to get hay fever from a palm? My next door neighbor says she thinks our sago palm is causing her sneezing. If that's possible, is there anything we can do about it, short of removing the tree?

A. Pollen from the male sago palm — really a cycad *(Cycas circinales)* — bothers some people. It is produced in the large, globelike brown center of the plant and gives off what many consider a foul odor during its development.

The female sago is a problem plant, too. The mealy white kernels of the seeds are very toxic if eaten. Sagos enjoyed considerable popularity in South Florida several years ago and plants purchased at that time are now mature enough to produce the fruiting bodies. Several people have successfully removed this flowering portion by cutting it off at its base with a sharp knife. Removal of flowers or fruit will not kill the plant.

Q. Recently a friend gave us a sago palm. After we planted it, the palm has been declining and the fronds droop to the ground. Can this be from the shock of moving? What about soil?

A. All plants suffer some degree of shock after transplanting. In this case, the fronds should have been removed before transplanting. Soil structure, if within normal limits, should not be a factor.

Q. I would like to know if there is any value in painting the lower portion of palm trees white and, if so, what sort of paint should be used.

A. In colder climates, paint was once the most effective deterrent against boring insects on fruit trees like apples, peaches and pears. (The lime in the paint discouraged them). After insecticides like Dieldrin and Endrin came along, its use dropped off.

Florida citrus growers have used paint to protect trunks against sun scorching after a heavy canopy of shade has been removed. This occurs in "top-working," when trees are cut back to the stumps and a grapefruit, for instance, is changed over to a Valencia by grafting the introduction onto the grapefruit rootstock.

There is absolutely no advantage to painting palms in the tropics, however, since the Northern borers aren't a problem here. And it's the quickest way to turn a beautiful tree into a blight on the landscape.

6. Fruits

*For special index to this chapter,
see back of book.*

More different kinds of fruit will grow in Florida than in any other state in the nation, and new residents are quick to take advantage of the wide choice that can provide fresh fruit year-round. Homeowners can choose from dozens of varieties of oranges, grapefruit, and other citrus, and avocados — the most popular — through delicacies from around the world, including lychees, mangos, persimmons, jaboticabas, carambolas, and sapotes.

Floridians ask more questions about fruit trees than any other subject — most related to the intricate and sometimes unpredictable fruiting mechanism of the over 300 different types known in Florida, but also traceable to the pests and problems that are attracted to them.

But because many trees almost take care of themselves, growers may overlook the fact that with regular examination they can help avoid puzzling drops in production or quality of fruit. Maintaining adequate moisture in times of drought, keeping a copper fungicide on hand to protect against fungus diseases, and using nutritional sprays are three major areas of care needed by fruit trees here. Problems that can't be met with these basics, plus periodic fertilizing, almost always can be squelched with ease if caught early through regular examination of trees.

Q. I'm interested in planting fruit trees for the fruit, but my husband wants some that are ornamental, too. Which would be both attractive and good-tasting?

A. It's hard to narrow the selection since so many tropical and subtropical fruit trees make striking shade trees when out of fruit and delightful flowering trees when in bloom.

Family fruit tree picking is fun — and this prolific lychee tree can use plenty of hands. Lychees are eaten fresh here. Their flesh has a grapelike appearance and taste. They usually are alternate-year bearers and like applications of nutritional spray as well as ground applications of chelated iron a couple of times a year.

The Barbados cherry most often is grown as a shrub, but it makes a charming small tree, bearing white or pink flowers with fancy scalloped edges. It has a long blooming period and the sweetish fruit is very high in vitamin C. It grows to about 12 feet high.

Calamondins and kumquats are decorated with small, bright orange fruit almost year-round, but you have to be a jam-maker to appreciate their flavors.

The sapodilla *(Manilkara zapota)* makes an outstanding shade tree. Its sticky white latex yields the chicle used in chewing gum. The fruit tastes like a cinnamon-dusted pear.

The mamey sapote *(Mammea americana)* is tall, dark, and handsome with lustrous, thick, dark leaves and fragrant white flowers which can be distilled to make a liqueur.

Leaves of the satinleaf *(Chrysophyllum oliviforme)* are glossy green above, coppery reddish-brown underneath. Its sister, the star apple *(C. cainito),* has gold-lined leaves. Try the annonas, too.

The waxy, five-angled fruit of the carambola *(Averrhoa carambola)* makes this tree a "must" as an ornamental.

Q. What is causing my mango's leaves to die back from the tips? Some are brown along the edges, too. If my house plants were to do this, I'd think something was wrong with the roots. What about the mangos?

A. There could be a problem in the soil — salt toxicity or over-fertilization, the presence of excessive animal waste or soil contamination resulting from spillage or dumping of bleaches, acids, herbicides and other toxic materials. If you think any of these might be the cause, leaching the soil with continuous applications of water will help.

Three other possibilities are anthracnose (control with a fungicide), bark scale infestation (use a scalicide), or spray burn. (Apply sprays carefully, according to instructions.)

Q. What is ailing my mango tree? The symptoms are (1) the leaves seem to turn brown and die away, (2) the ends of the limbs turn black and rot off, (3) the trunk seems to have a black fungus on the bark.

A. The first symptoms suggest possible salt burn (if lightning damage is impossible) and the last could be sooty mold. If you live near a source of salt water, the well water you use for irrigation might contain enough salt to cause damage. Fertilizer applied to dry soil also could cause this type of damage. The pattern of damage would show the tips and edges of the leaves turning brown first. Insects usually would show a more irregular pattern of damage in the center.

You could have the well water checked for salt content. The sooty mold would be an indication of scale insects. Usually one thorough spraying with oil emulsion in June or July will control these pests. The mold grows on a sticky substance produced by the scale. With scales controlled, the mold should disappear.

Q. Could you give me the maturing time "from blossom to table" for Haden, Springfels, Kent, and Carrie mangos?

A. The time required for a mango crop to develop from blossom to mature fruit is about six months but it can vary noticeably due to different weather conditions from season to season. The trees may bloom from December into March, depending on the kind.

The prime picking season for the Haden and Carrie fruits is June, but some may ripen in late May. Springfels and Kent are midseason varieties with the harvest coming in July and August.

Q. My mango is spewing little golf ball sized fruit over my lawn. What could the problem be? I water, spray, and fertilize on schedule.

A. The "golf balls" are partially a result of lengthy cold weather which made pollen less vigorous and, in turn, produced less robust fruit, many of which are seedless or have damaged seed. You'll still get tasty fruit this year, but the crop and the size will be smaller.

Drought usually results in smaller fruit, too.

Q. I understand you can graft a mango or an avocado with different varieties of mangos or avocados and be able to have fruit almost year round. I'd like to be able to do this myself. How can an amateur pick up the technique?

A. Keep your eyes open for workshops and programs on grafting sponsored by the County Extension Service and clubs like the Rare Fruit Council International. But meanwhile, contact your County Cooperative Extension Service and ask it to send you the free booklets "How to Grow Your Own Mango Tree" and "How to Grow Your Own Avocado Tree." These illustrated publications tell you how to grow stocks, collect budwood, graft, and care for your grafts.

Q. Are some mangos more prone to cracking open than others, and does anthracnose cause the splitting?

A. Andersons and Springfels have a strong tendency to split as they mature and Irwins and Hadens also will open occasionally. Prime cause is climactic; heavy rains following extended periods of dry weather result in fruit-splitting pressures within the mango. But, again, anthracnose can cause it, too. The black spots are dead areas on the skin which don't expand as the fruit grows.

A drought can cause soluble solids to build up to high levels. Then when rain finally comes, these solids, mainly carbohydrates, quickly absorbed the fresh supply of water. So did the tree as a whole and, trying to achieve equilibrium, the water diffused rapidly into the fruits to hold the carbohydrates in solution. Internal pressure rose to high levels, and the result: fruit splitting.

The hard-hit mangoes may also be troubled with anthracnose, which is aggravated by high humidity, heavy rain or dew during the critical period in which infection develops. Preventive spraying with neutral copper is recommended to help avoid these large black spots on fruit. Benlate or Benomyl may be used alternately with copper.

Use a nutritional spray, also, to give the tree adequate zinc.

Q. My mango has grayish white and pale green discolorations up and down the trunk. What are they and how can I get rid of them?

A. They are tiny plants called lichens and they also come in pink and red. While lichens themselves are harmless, they thrive on trunks that are getting more sun than a normal tree should. A thin leaf canopy perhaps should be your real concern. Too few leaves can be the result of nutritional deficiencies, scale and mite infestations or defoliation by disease. Check for these problems. There is no need to control the lichens.

Q. The new leaves on my mango are coming out smaller and smaller, and their shape is distorted. They don't improve with age, either — they're leathery and narrow. I apply fertilizers regularly and I can't see any bugs on them. Can you give me some help?

A. These sound like symptoms of zinc deficiency. You should be applying complete nutritional sprays (containing zinc) to the leaves of your mango at least once a year to prevent these signs of diet deficiency. After the symptoms appear, zinc deficiency may be corrected by one or two sprayings.

Q. Every year I spray with neutral copper to control spotting on the leaves and fruit and every year I still have to throw a lot of my mangos out. What could I be doing wrong?

A. Anthracnose is a big bugaboo with mango growers who are raising some of the older, very spot-prone varieties like Haden and Carrie. You'd have much less of a problem with one of the newer mangos such as Tommy Atkins.

But in any case, it's vital that you not only make your spray applications at tho right time but thoroughly. Successful control of fungus diseases calls for all susceptible parts of the tree to be coated with the fungicide before infestation starts. Sprays applied after infection (which occurs several days before the disease shows itself) won't help the existing problem, although it will curb its spread.

Neutral copper must be reapplied as new tissues become exposed by growth and as spray residues are reduced by weathering.

Spraying should begin when bloom spikes are three to four inches long. Benlate and copper may be alternated.

With most mango disease problems, spraying should be done immediately following wet weather but after the tree has dried off.

Q. We had beautiful mango fruit this year, but all the fruit was rotten around the seed. I would like to know what to do and when to do it.

A. This may be a condition known as "jelly seed." The area around the seed apparently matures ahead of schedule and then starts to break down. There are suspicions that too much nitrogen is the cause. Another theory is that the tree may have sent its roots into a septic tank drainfield.

All mangos tend to ripen from the seed outward, some more rapidly than others. These may have to be picked earlier.

Buy a fertilizer with a potash level (the last number on the fertilizer bag) slightly higher than the nitrogen level (the first number). Magnesium also helps prevent "softnose," (another name for this condition). In acid soils, raise the level of calcium and magnesium.

Q. My Brooks mango is producing in great quantities but the fruit is dropping prematurely. Black areas form on the mangos and the spots seem to be infested with small round insects and worms. Oil emulsion, nutritional, copper fungi, and regular fruit sprays have not helped. What more can I do?

A. The black spots are anthracnose. Control by continued spraying with a neutral copper fungicide. The insects and worms could be scavengers, larvae of Caribbean or papaya fruit fly.

Q. My Sensation mango has always produced good fruit. This year the fruit looked good but a couple of weeks before normal ripening, the mangos became soft at the bottom and rotted. They had a rotten odor when cut. More than 100 were ruined. Anything I can do to have better luck next year?

A. This very likely could be a fertilizer imbalance where the tree is getting too much nitrogen or potassium — probably nitrogen. It is called "softnose." Cut back on all fertilizer if you suspect this problem. This may also be disease or even a minor element deficiency. If you don't believe you've got "softnose," give more details to your county agent.

Q. My mango tree blooms heavily, but when the fruit gets to be about the size of the end of my thumb, it all drops off. What causes this and what can I do about it?

A. The most common cause for such a problem would be an inadequate fertilizer program.

Unless a tree has been neglected, three fertilizer applications a year are recommended. A medium application can be scheduled when the fruit is about golf-ball size.

The heaviest application is made just after the crop is harvested. A medium or light application, depending on the condition of the tree, may be made in the fall when the rainy season ends.

If you have not made an application in quite a while, you can go ahead and fertilize the tree now.

Anthracnose and powdery mildew can cause premature drop, also.

There is a theory that low temperatures during the blooming period can be blamed for poor fruit set, but it's doubtful that, in your case, such damaging temperatures were experienced.

Q. My mango leaves are showing a V-shaped dieback from the tips. I usually think of root problems when the tips of leaves turn brown, but I'm giving my tree the same care I always have. What's wrong?

A. There are several possibilities. Among them is a hard-to-spot bark scale. There might also be scale on the leaves which could be treated with Malathion.

You might also suspect a nutritional deficiency — not enough potassium, magnesium, or calcium (not likely in Dade County). If you have a pool, chloride runoff from the decks could cause dieback. So could a bleach solution dripping from your roof if you cleaned it in this way recently.

Q. Can you tell me what the problem is with the enclosed leaves from my mangos? They're covered with tiny bugs and turning brown.

A. When the red-banded thrip is enjoying a good season on mangos it can cause the defoliation of a tree. They give leaves a brown, rasped look and can be found on both sides. Adult red-banded thrips are all black; the immature pests have a stripe across their back. They measure about 1/32 of an inch.

Prompt spraying with Malathion is the answer.

Q. Many of the new blossoms on my mango are coated with a grayish powdery-looking fuzz. What is this and what should I do about it?

A. Mango owners often notice this condition on trees coming into their second bloom. If not controlled in its early stages, powdery mildew, a serious problem in mango-growing Florida and in India, will affect stalks, cause young fruit to turn brown and shed, and give the tree a severely burned appearance. On mature fruit it appears as irregular blotches.

Spraying or dusting with sulphur a couple of times should curb it. You can combine the sulphur with the neutral copper application which should be made every two to three weeks for anthracnose (black spot). Sulphur and oil (for scale) should never be applied within three weeks of each other, incidentally. Oil is a material that shouldn't be used as a control when trees are under stress from drought, either.

Q. I heard that mango scab is the worst it's been in years. Would you describe it for me? My trees don't look so good and this might be what's bothering them.

A. It's natural for the scab to be prevalent in rainy springs. Generally a nursery-type disease that is found where plants are crowded and the air circulation is poor, mango scab doesn't do as well where trees are widely spaced.

On tender new growth it appears as dark brown to black, circular to angular spots from 1/25th to 1/16th inch in diameter. As it develops, leaves become crinkly and distorted and the tree sheds prematurely. On older leaves the spots appear grayish, surrounded by a narrow dark border, sometimes breaking into "shot holes."

If conditions remain the same and the disease is allowed to spread, serious defoliation and a reduction in fruit production could result. The remedy is spraying with two to three tablespoons of neutral copper (wettable powder) with a surfactant (sticker-spreader) added to help adhere the material to the leaves. Benlate may be used, also. Spraying should be done at three to six week intervals, depending on the weather.

Q. My mango hasn't bloomed so far this year and now it is coming out with new growth. Is there any way of knowing when it's going to bloom?

A. Probably next year. Kent varieties, especially, can be thrown off-balance by warm winter. Once you see new growth you can expect your tree to have bypassed its reproductive phase.

Q. Why aren't my mangos blooming? I've seen others around me in bloom, and I've given them good care.

A. Mangos can bloom very erratically as a result of an especially warm winter. Prolonged cold weather can also influence blooming.

Their performance then, more than ever, resembles the spotty behavior of mangos in the tropics where the summer-winter temperature variation is less extreme. One side or even one branch may be the only part of the tree blooming. The outcome is not predictable as there may still be a lot of bloom to come. Or none.

Q. I have a beautiful, bearing Springfels mango which puts out a lot of fruit, but when they're half mature, problems set in. Something bores the fruit, causing juice to ooze out, and that's the end of the fruit.

I spray faithfully with copper and Maneb. Could this be the fruit fly? My surrounding Keitt mangos are not affected.

A. Fruit flies seldom bother mangos, but if you're talking about a tiny hole, they might be suspected. A probability that fits your description a little better is stink bug or pumpkin bug damage, an infrequently encountered problem for which there is no approved control on mangos.

Woodpeckers are a serious problem to be suspected with holes from pencil size to larger. Other birds more exotic like the South American bulbul and wild parrots favor mangos, too. Squirrels also nibble on mangos but their holes look like bites.

Q. Both my mango and my lime are losing their leaves and I can't tell what's wrong with them. Is this something seasonal or should I be doing something about it?

A. It's normal for trees infected with spider mites. They usually cause the most damage in dry weather. Spider mites on citrus, or avocado red mites on mangos, feed on the upper surface of the leaves, making them turn brown and drop. Spraying with Kelthane or sulphur will control these pests.

You might also keep your eyes open for a quartet of unsavory fungi — melanose, scab, greasy spot, and red alga — that are problems on all citrus. Spray with copper or Ferbam at the beginning and end of bloom but avoid full-bloom treatment. Pre-bloom takes care of the scab, a post-bloom spray goes after the melanose, and a summer spraying should take care of the greasy spot. Apply all of these and you should have no problem with the red alga.

Q. Our citrus trees are being bothered by whiteflies and the leaves look black and sooty. What can we do?

A. Both conditions can be cleared up by the application of Malathion.

The whiteflies produce a sugary material upon which the sooty mold feeds. After the spraying — which should be thorough, covering all parts of the trees — don't be impatient for the black material to go away. Very often it takes several months, depending upon the amount of rain. It won't hurt the leaves or the fruit — it is only unsightly.

Q. How can I tell if my trees have the citrus blackfly, and, if they do, what should I use to get rid of it?

A. Easiest to spot are the eggs of the blackfly, which typically are laid in a spiral pattern on the undersides of the leaves. The blackfly itself has a bright, brick-red head and thorax as a young adult, then becomes covered with a fine powder that gives it a generally slate-blue appearance — hence the name "bluefly" in the Bahamas.

The fruit fly, a pest throughout the world tropics, is especially obnoxious on citrus but also infests mangos, ardisia, avocados, pomegranates, sapodilla, cashew, rose apple, sugar apple, soursop, sapote, breadfruit, star apple, guava, mamoncillo, canistel, and many other fruits. Malathion, licensed for citrus, is the only approved control.

Q. I have two citrus trees, both with trunk problems. The bark at their base just above the ground has been peeling off, exposing the wood. I've tried wiring it back on but it doesn't "take." What's the matter?

A. "Foot rot." It can be caused by mulching citrus trees (don't), a build-up of grass around the "foot," or wounding the trunk with a mower, hoe, or sprinkler. "Foot rot" can also cause chlorosis, loss of leaves, wilting, and reduction in yield and fruit size.

Moisture from irrigation or rainfall plus poor air circulation around the crown initiates the disease, so clean up this area, removing grass by hand. Then clear the trunk of diseased tissue (any bark stained tan to yellow on its inner surface — and don't be afraid to take disease-free tissue with it to be sure), and spray with a neutral copper solution. Planting trees too deeply is another cause of "foot" problems, also called "crown rot."

Q. I can't say much for my citrus this year. The fruit is dry, pithy, and insipid, although it seemed to develop normally and the leaves are a deep green. There's very little juice. In fact, there's very little fruit. What caused this?

A. This condition is most often found either on very young trees that are in robust growing condition or on older trees which haven't been watered or fertilized properly.

Try putting your tree in a more even state of growth by watering thoroughly about once a week in cool months and not more than twice a week in the summer. Fertilizer should be applied three times a year — in spring, midsummer, and fall — at a rate of a pound for each foot of canopy diameter. Spread it uniformly over the entire root area and be sure it's washed thoroughly into the soil.

A good fertilizer to help meet this pithy problem is a 6-6-6 with 3 percent magnesium. Chelated iron should be applied to the soil, and one or more applications of a nutritional spray containing copper and the leaf-greening "minor" elements is helpful.

Q. Please tell me what kind of paint should be used on stumps of citrus trees after cutting off dead and large limbs? I have grapefruit, tangelo, and calamondin. Do they require different kinds of paint?

A. All leading garden supply stores have special pruning paint available. Always use a water base paint. If you have cut away much of the foliage, it will also be a good idea to whitewash the limbs to prevent sunburning. Sun-dried bark overheats and kills underlying tissue, resulting in extensive dieback.

Q. What could cause a half-juicy orange? My Temples this year are large and succulent except near the stem end where the fruit is dry and woody.

A. Your tree will grow the fruit for you but no orange has been developed yet that will harvest itself, which is often the problem with partially woody fruit. Waiting too long to pick the fruit after it matures gives it time to begin drying out.

Q. We have two orange trees in our backyard that are about five years old. We have a good crop every year, but we never get to eat them. They are always dry inside with no juice. They never turned orange, and this year is no exception. What can we do to remedy this situation?

A. It would help to know the variety of oranges you are growing and some more details about the trees. Researchers report some kinds tend to have dry fruit if grafted or budded on the wrong rootstock.

Since that information isn't available, another thing to consider is the possibility of overwatering. Fruit may be dry if water is abundant. This situation often comes up where trees get water from lawn sprinkler systems. Grass needs water more often than do orange trees.

Oranges should be used when they are at a peak of quality. Since the fruit on your trees doesn't change color, it may be that you are missing this peak, and the fruit is drying out on the tree.

Too much nitrogen could contribute to the lack of coloring in the fruit, although oranges don't usually color as well in the Miami area as they do in the main citrus area in the central part of the peninsula.

Q. My orange tree is loaded with fruit but it won't turn orange. I water and fertilize faithfully. New buds are beginning to form now. Should I pick all the fruit off the tree?

A. There's a possibility that the fertility level is too high, especially if your tree is located near or over a septic tank drain field. In a season when there has been enough cold weather, there will be more orange pigments. However, just because your fruit is still green does not necessarily mean it is not ripe nor good. Many Florida mid-season and Valencia oranges are green but are fully ripe when cut open. There is no need to pick your fruit just because buds are beginning to form.

Q. What can be done for our orange tree? The oranges are split and full of the fruit fly. Some leaves are black and curled. We water it well and fertilize about three times a year.

A. Splitting of Valencias and sweet seedlings is often common in the fall and, of course, provides an attractant to flies and other insects. It is too late to correct this year. Don't drown the tree. Use a good balanced fertilizer, probably a 6-6-6 mix with three percent magnesium. Also use a neutral copper spray with manganese and zinc about three times a year.

Q. My orange tree has several leaves that are covered with a black film. I seem to remember having this trouble before but, anyway, what is it, will it spread and what can I do about it?

A. Your tree is experiencing sooty mold. The mold is a saprophytic fungus, always present in the air, which can establish itself quickly when it comes into contact with the "honeydew" secretions of sucking insects such as aphids, immature whiteflies, mealybugs, and soft scales.

Sooty mold takes nothing from the plant but (in severe cases) may interfere with its photosynthesis by shading out the sun. Oil sprays will help correct this condition and remove the mold. (Do not use if temperatures drop below the 50s or if leaves are wilted.)

Meanwhile, you should go after the insects that are causing the problem. Aphids multiply like rabbits, and you have to look closely to spot the tiny green bugs. One signal that they're present is a stream of ants running up and down the limbs. Ants will carry aphids piggyback and they feed on the sweet honeydew secreted by the insects.

Aphids, scale, and whiteflies not only encourage the formation of sooty mold but can cause "invisible" sucking damage. It may not become apparent until next spring, since new shoots arise after the terminal ones are injured. Control these bugs by applying a Malathion spray.

Q. What could be causing splitting bark on my orange trees?
A. Growth cracks on citrus and other fruit trees are normally a result of excessive water. When trees are watered more than enough or when there is a period of very heavy rain, trees will absorb more moisture than they can use and the result is splitting bark. Growth cracks are dry.

If the crack is infected with fungus, the cause is "foot rot" (see p. 102). Clean it out, spray it with neutral copper, let dry, then coat it with pruning paint. Do not attempt to fill it with concrete until the wound is thoroughly dry. (Concrete is not necessary.)

If the tree is not in good growing condition, a nutritional spray or light fertilizing should be applied to speed recovery.

Q. A three-year-old tangelo tree, faithfully watered, fertilized, and sprayed with nutritional spray, blossoms and sets tiny fruit, which keep dropping off. The tree appears to be healthy but something is chewing at the leaves, which are curled. Occasionally I notice a white filmy substance on the curled leaves.
A. Sounds like aphids have gotten to the tree. If this has been a continuing situation, considering the young age of the tree, it has probably been completely stunted. Spray with Malathion. Since the tree blossoms and tries to set fruit it may also have root problems. Remove weeds around the trunk, all grass, and mulch from the immediate area. Be careful not to waterlog the tree.

Q. My grapefruit tree is not too old but a hurricane blew most of the leaves off. Last summer it blossomed, but only about 12 or so fruit were on the tree. It blossomed again, and it has a lot more fruit. I cut open one of the second bloom fruits and instead of pink it was white and awfully sour. What's wrong with it?
A. It isn't unusual for a grapefruit tree to produce two blooms in a single season. The trees often bloom and set fruit in spring and bloom again in June, setting additional fruit then.

Fruit from the two blooms will not mature at the same time. So, it may be that the fruit you sampled just hasn't had time to mature properly.

To account for the color change, the fruit you picked might be a fluke, and the other fruit may be normal.

Another possibility is you have a grafted tree on grapefruit rootstock. A branch or shoot growing from below the union might produce white fruit.

Q. For no reason I can see my Key lime has several branches that are dying back. What can I do before I lose the tree?

A. First consideration is root damage by nearby construction, weed killers or toxic chemicals such as chlorides. Then examine your tree's nutritional program. Could you be overfertilizing, overwatering, or underwatering?

Some very tropical citrus like Key limes and Isle of Pines grapefruit seedlings tend to defoliate seasonally, too.

Citrus trees are occasionally subject to a dieback thought to be caused by a virus but possibly a fungus. Not much work has been done on the subject by researchers so there isn't much that can be offered in the way of remedy. Cutting out the diseased wood helps little.

There is a virus or fungus that attacks Key limes with fatal results, but before you give up on a declining Key lime, apply a nutritional spray to the leaves and magnesium to the ground.

Q. Just before my grapefruit tree started to bloom I sprayed it with Malathion. At least I thought it was about time for it to bloom, but nothing happened and this year I have had no fruit. When we bought the house we had a wonderful crop. Same last year. Could the Malathion have stopped it from blooming?

A. This phenomenon occurs with grapefruit, especially Duncans. They will fruit several years in succession, then rest. There is little you can do about it except perhaps to remove excess fruit in good years, helping to conserve its energy and keep it on an even keel.

Q. I'm having trouble with my grapefruit tree. The bark splits on the trunk and a gluelike substance comes out. What could I do to remedy this situation?

A. That clear, amber colored goo is either the result of gummosis or virus infection, and it is thought weed killer can sometimes promote it. There's no remedy; spray with neutral copper against other infections.

Q. I think the former owners of our property were mistaken when they said the little tree in our rear yard was a kumquat. It is my understanding that kumquats do very well here but so far our tree has produced only a handful or so of little round tart fruits. A neighbor suggested that our unusual weather this year might be responsible. Is he right?

A. Indirectly. Both the round 'Marumi' kumquat *(Fortunella japonica swingle)* and the round 'Miewa' *(Fortunella crassifolia swingle)* need a cooler area than South Florida for good growth and fruiting. The most successful type for this area is the oval kumquat, *Fortunella margarita swingle,* commonly called the 'Nagami.'

Nutritional sprays are essential for the best kumquats.

Q. Something's wrong with the blossom end of the fruit I'm getting from my calamondins. It starts out as a water-soaked, grayish-tan area and spreads. Eventually the flesh gets squishy. Could this be from too much moisture or what?

A. Limes, lemons, and calamondins in the summer and early fall are sometimes victims of stylar-end rot. It occurs most often during periods of hot, humid weather and heavy rains. The cause is being researched, and when the answer is found it will be released by county agents.

Q. My Key lime tree, planted more than a year ago, has not increased in size. The few new leaves have been very dark green, small, and wrinkled. Everything else on my property has flourished. I have about decided to remove the tree but thought I would first ask for advice.

A. It is true that Key limes respond better to "undercare." The leaf wrinkling could be caused by aphids or some other sucking insect. The possibility exists that you may have planted the tree too deep. Others are too much salt, overfertilization, etc. I suggest you give this tree a nutritional spray to provide minor elements. A Malathion spray will knock out the aphids. Several water flushes will reduce the salt potential if it exists.

Q. I am enclosing a leaf from my Key lime tree. Can you please tell me what this fungus is, and if there is a cure? It has spread to other small trees.

A. The leaf shows the typical combination of scale insects and sooty mold. The bottom side is dotted with the scales which appear as flecks of wax. These pests suck juices from the leaf and give off a sticky substance, which is an ideal place for the sooty mold to grow. The mold isn't harmful to the tree unless it becomes so dense that it cuts off sunlight to the leaves.

A Malathion spray should control the scale, and once they are dead, the mold should disappear over a period of time. Don't be too impatient, because it will take time, rain, and wind to remove the mold.

Aim for thorough coverage when you spray the tree. Since the scale usually stay on the undersides of leaves, direct the spray upward to hit this area. Enough spray will fall back to coat the top sides.

Q. I'd like to know what formula of fertilizer to use to make a lime tree bloom. Last year I transplanted a Key lime tree from a container. The

tree is green and now is head high. When I transplanted the tree, it was full of bloom. The bloom dropped off and it has not put on bloom since.

A. You will have to wait until the tree is ready to bloom again. In the meantime, avoid the temptation to overfertilize it.

Even though a fruit tree may bloom and bear fruit while it is in a container, chances are it will go a season or more, once it is in the ground, before it will bloom again. But it is growing a large root system and increasing the size of its top. When the new growth is mature enough, the tree will bloom and fruit.

There is a temptation to fertilize a tree heavily in the hope that it will bear sooner. But a Key lime needs only half as much fertilizer as other citrus trees. Given too much, it is prone to develop a fungus problem that causes dieback of the twigs.

Since your tree is well established, two applications of a general purpose garden fertilizer a year should do. It's wise to use one with added magnesium. Measure the tree from branch tip to branch tip. A half pound of a 6-6-6 fertilizer for each foot of spread will be enough. Don't forget nutritional sprays, either.

Q. I'm enclosing some little black spots I have all over my Key limes. They look to me like scale but I'll let you tell me what they are. Are they harmful to my tree? The tree looks healthy enough and bears well. Should I spray to get rid of them?

A. Yes, they are scale, and are called, appropriately enough, black scale, although other black scale insects have other names. The black scale is identified by the ridges forming an "H" on its back. Malathion will get rid of them.

Scale on citrus can become a mess, prompting the development of black, filmy sooty mold on the leaves and sapping nutrients from the tree. If you spray now you won't need to stage a full-scale war later.

Q. I don't know what's causing the chewed leaves on my Persian limes, but nothing I've used as a spray has had any effect on them. I saw two "orange dog" caterpillars on the trees. Could these be the problem?

A. If the damage is being caused by the "orange dog" (which bears an unfortunate resemblance to bird feces), best thing to do is simply pick them off. They won't bite or sting. With bad infestations, use Malathion or Diazinon. "Orange dogs" are the immature stage of swallowtail butterflies.

Check for other bugs, too. The katydid is almost invisible since it looks like a leaf eating a leaf. Hand-picking is also advised; spray with the same materials for an infestation.

Another insect which damages leaves is the leaf-cutting bee. It leaves very symmetrical, circular notches. It is seldom present in numbers large enough to worry about.

If you are unfortunate enough to have citrus root weevils, here's how to check for them: spread a light colored tarp under your tree and shake the tree. The pests will drop, giving you an idea of how bad an infestation you have.

Citrus root weevils are about a half-inch in length and a metallic green in color, but are hard to spot on a tree because they "play 'possom" and fall off when the leaves are disturbed. They are not very harmful to the foliage, other than the unattractive appearance of the leaves, but their larvae in the ground can be a serious problem. Soil drenching with chlordane is the only answer.

Q. Can you tell my why tiny limes keep falling off our tree? It is healthy and loaded with blooms but none mature, falling off after getting about one-fourth inch in diameter.

A. Since you don't mention any sign of insect damage, the answer probably is in the fertilizer program. A regular program would include four applications a year of a 6-6-6 fertilizer which also includes three percent magnesium. Or you could use somewhat ligher applications six times a year.

Be sure to make one or two applications of a nutritional spray during the year. With adequate fertilizer your tree should produce limes.

Q. The leaves on my lime tree turned yellow and fell off. It is bearing plenty of fruit but is beginning to look like a skeleton. What is wrong?

A. Check the tree for foot rot. Remove mulch, weeds, soil, or debris from the trunk. The tree may already have lesions. If not checked, foot rot can completely girdle the trunk and kill the tree.

The condition also could be "greasy spot" on the leaves. It looks like it sounds and is most easily avoided with preventative sprayings of neutral copper.

Q. What could be the matter with my lime and lemon trees? The branches suddenly are covered with blotches of short red fuzz.

A. It's the fruiting structures of red alga that made the blotches appear — and failure to spray at least once a year with copper. Once the alga is on the scene it can cause the bark to crack. Spraying them will not get all the alga inside the cracks, but if you apply a couple or more sprays a year, you'll eventually have it under control.

Q. Why would my lemon tree (very small) lose all its leaves, yet blossom and begin to grow very tiny fruit?

A. The loss of leaves could be due to mites, which can be controlled with a miticide such as Kelthane. Another possibility is "greasy spot," a fungus disease. A neutral copper spray would be the way to fight this. The copper can be applied three times a year.

The most likely cause is "foot rot," resulting from planting the tree too deeply, piling soil or other material around its base or crown, or letting sprinklers soak the trunk. Loose or splitting bark is the tip-off. Remove it, clean the area, and coat it with neutral copper. Use spray-on (preferably) tree wound paint after the copper has dried.

Q. What I want to know is how to raise an avocado seed in water until it develops roots, and how long to keep it in water before planting.

A. You can start your avocado project by inserting at least three toothpicks into the side of the pit. Space the picks so you can suspend the seed in the mouth of a glass or other container.

The slightly concave end of the seed should be down in the water. There usually will be a slight point on the opposite end.

The seed should be suspended to allow about half of its surface to be under water. You'll be able to observe the roots growing in the water.

When the seed is well sprouted, it can be transferred to a container of soil. A good medium would be half peat moss, a quarter perlite, and a quarter soil.

Q. My problem is cracked avocados. Every one is cracked and partially rotted before maturing. Can you explain this?

A. This most often comes about when anthracnose spores get into the fruit as a result of injury or scab lesions. The tree should be sprayed with a good neutral copper. Late blooming varieties are more susceptible. Spray about once a month.

Q. We have an avocado tree in our backyard that oozes a dark brown "goo" from a small hole in the trunk. How do we go about repairing it?

A. Clean the hole, removing all decayed wood, then paint the inside with pruning paint. Be sure the oozing has stopped. The hole is next filled with concrete but not past the point where the bark meets the wood. This will give the bark a chance to grow over the concrete, eventually hiding it.

Q. I recently planted a young (four-foot) Pollock avocado tree. Would you tell me when this tree blooms, when the fruit matures, its feeding and care.

A. When your young tree is mature enough, you can expect it to bloom in late winter. The time may vary a little, depending on the weather.

The Pollock is an early variety that matures in July. Its care resembles that of other avocados.

During the first year the tree is in the ground, you can fertilize it monthly for quick growth. Start with a light application (say, a small handful) and increase the amount a little each time. At the end of a year, the application can be about a pound.

Gradually space out the applications over the next year or two, increasing the amount each time. When large, the tree may need only a couple of applications a year — in the fall and in late spring.

During the first year, plan to make three applications of nutritional spray. After that, once a year, in January, should do.

Some gardeners never spray their avocado trees and still harvest reasonably good crops. But you may find scale insects on the undersides of leaves. Malathion will control them.

Mites can discolor foliage and you may want to use a miticide particularly during the winter months. Scab disease may cause blemishes on the skin of the fruit, but doesn't affect the eating.

Q. Approximately four years ago I planted a seed from an avocado purchased at a grocery store. It has reached a height of about 10 feet and is very slender. Its leaves look pretty healthy but I wonder when it will spread and if it will bear fruit. We haven't given it any special care, other than fertilizer two or three times.

A. Seedling citrus, mangos, and avocados are far more unpredictable than the weather. When allowed to grow from seed, rather than being grafted with budwood from an older tree, these trees tend to be less spreading, growing tall and erect with what is known as "seedling vigor." While they are still in the vegetative stage (not yet producing fruit) they will put on a tremendous amount of growth, similar to childhood and adolescence in a human.

After finally passing through a fruit tree's version of puberty, they will begin setting fruit and become more spreading and slow-growing. "Finally" may be a period of time as short as three years and as long as 10 or 20. A seedling's average time to bearing is seven or eight years. This is one of the reasons why people who can't stand long lines or suspense buy grafted trees. The budwood grafted onto the seedling stock has already gone into the reproductive phase, making the branches it produces virtually as "old," or in the same spreading, bearing stage, as the tree from which the graft was taken.

Another reason why grafted trees are preferred is that their fruit will be of the same high quality as the "parent" tree while, with a seedling, the fruit may be bad or good; abundant or skimpy in number.

There is nothing you can do to hurry up a seedling tree, so relax and enjoy its leaves — or find someone who can graft it for you.

Q. The first year I had my avocado (a Pollock) it bore three fruit. That was four years ago, and until this year I didn't see any more. I was told to spray with neutral copper and I did, several times. The result was one mature fruit and several tiny ones which fell off. I had fertilized several times with a 6-6-6 and expected more, to say the least. Could I have hurt the tree by digging in some vegetable peelings last winter?

A. Probably not, if you dug out from the base of the tree so the material was well decomposed by the time it reached the roots.

It is believed, although not conclusively proven, that Pollocks have a built-in birth-control system that you may have encountered. Also, they may tend to be "self-infertile."

Avocados have a bizarre approach to fruit production. The varieties are divided into "A's" and "B's." Pollock is a "B," which means that its flowers open in the afternoon as females, close, then open the next morning as males. The "A" varieties open in the morning as females, close about noon, then open the following noon as males.

Not all the blossoms on a tree, of course, open the same day, but it is only during that part of the day when they are open as females that they can be pollinated.

Some avocados are self-pollinating because their flowers don't open and shut like clockwork. There is an overlap. But it is thought that the Pollock, especially, is an avocado that needs another pollen source — an "A" variety like the Lula, which blooms at the same time of year. It should be planted within 20 or 25 feet of the Pollock because, although bees may travel great distances to reach pollen sources, they haven't the capacity to lug pollen from much farther away than that.

Q. How come I get any fruit at all on a Pollock if it's thought the opening of the male and female blossoms don't overlap?

A. A solitary Pollock may bear some fruit due to an idiosyncrasy peculiar to this variety.

Pollocks are famous for "parthnocarpically" setting fruit. This means they can set fruit asexually. Pollination takes place but no fertilization. Even a grain of sand instead of pollen can trigger the regulators that will cause fruit to start forming.

Depending on the vigor of the tree and other factors, all they may result will be little seedless cucumber-shaped fruit that will eventually fall off, or a few fruit that may actually reach maturity.

So, if you own a Pollock, buy a Lula — the taste is worth the effort.

Q. My avocado is beginning to bloom but all the leaves are dropping. Except for the bloom, the tree looks like it is dying. What could be the matter?

A. Depending on the condition of the tree, about bloom time it is normal for many avocados to lose their leaves, particularly West Indian varieties like Pollock, Waldin, Fuchs, and Simmonds. In times of unusual cold and drought, all types may drop their leaves.

Q. The hurricane of two years ago broke off an avocado tree. Since that time it is nicely branched, but has not had blossoms or fruit. Can I ever expect fruit again?

A. Yes, given enough time, the tree should produce fruit again. In many fruit-bearing trees, the growth must reach a certain maturity before it is possible for it to flower and grow fruit. Yours may bloom in the next few months.

In the meantime, keep the tree in good condition with regular fertilizer applications, irrigation when necessary and pest control.

Q. I have a Cavendish banana plant in my yard that I planted about three years ago. The plant so far has had at different times three stalks of bananas, about 30 separate bananas to a stalk. They look good, but they do not develop or fill out. After a while they just start rotting. The leaves are beautiful — healthy and very flush. I fertilize and water regularly.

A. A banana plant will use large quantities of fertilizer and water. Lack of some element may cause the problem you mention. These plants can be fertilized monthly and they will benefit from a layer of mulch. The mulch should be renewed as it breaks down.

An application of a nutritional spray to the leaves also may help. Weekly watering should be sufficient in a fairly well drained spot. Banana plants may look good, even when growing under rather poor conditions, but they need plenty of fertilizer to produce the large bunches a Cavendish is capable of giving.

Q. What causes the splitting of my banana leaves? I give it plenty of water and fertilizer every two weeks. It is about two years old and has never borne fruit.

A. Wind is the cause of the splitting. That's why bananas should be planted where they have some protection from wind if possible.

The splitting may not make a neat appearance but tattered banana plants seem to grow and fruit well anyhow.

The length of time required for a plant to produce fruit varies, but many strong young plants will bloom in a year or a little longer. You might try a nutritional spray on yours. A heavy mulch also is helpful.

Q. I have a bunch of banana questions. How old does a banana have to be before it will set fruit; should you remove the suckers when they appear and the flower, too, after the fruit sets; and is it necessary to cut down the stalk after you've picked your bananas? Also, what about fertilizers?

A. Bananas take 12 to 15 months to flower and about three months after that to mature fruit. (They're mature when their angles become rounded; pick them then and let them ripen off the plant.)

Suckers appear before the flowering stem shows itself and it's best to remove all but one. This will throw strength into the flowering stalk and the plant that will take its place after fruiting. The suckers can be transplanted. Removing the flower after the fruit has set may increase the size of fruit.

After the fruit is harvested, the stalk should be cut down, chopped up and placed around the suckers as mulch.

Speaking of mulch, bananas thrive on a heavy layer, constantly renewed. They also like frequent, light fertilizing with 6-6-6 and an annual application of a nutritional spray to the foliage. An ideal soil is deep, fairly heavy, and moist but well drained. They will, however, settle for less.

Q. Can you tell me why my banana trees are so unhappy? We set them out a year ago in a good hole with muck and sheep manure, and we have fertilized them about three times. They have not grown one bit bigger in that year.

Nothing seems to be eating them, but they look sick and puny. What should I use for mulch and what for fertilizer? Nutritional spray?

A. Bananas can use a rich diet. Three applications of fertilizer in a year would be rather light.

Why not try fertilizing your plants once a month? A general purpose garden fertilizer should do a good job.

Also give the plants a nutritional spray to provide minor elements. This will be the standard spray sold in all garden supply stores.

Bananas also want plenty of water. Soak the root zone weekly unless rain does the job for you.

A heavy mulch, renewed as the old material decays, often gives a good boost to bananas. It keeps the soil from drying quickly and it adds organic material to the soil.

If the plants don't respond to this program over a period of two or three months, you might replace them with vigorous, well-rooted suckers and try again.

Q. We have a banana patch that gets bigger and bigger but we seem to be getting smaller and smaller bananas. I've tried to get my husband to thin out the plants but he says he likes the lush tropical look. I like the fruit even more. Any suggestions?

A. Thinning bananas does seem to affect their growth and production of "hands," and you can increase the size of the fruit on individual plants by removing, say, all but three hands. From three to five hands are about all the average banana gourmand can eat, anyway.

Q. Very few banana trees in my area of the Keys produce more than a third of a stalk of bananas. If this condition is due to lack of fertilizer, what kind should be used and how often?

A. It may be that these poor-bearing bananas are lady-fingers, and if so, they are notorious for this behavior, often showing no more than three to five hands per stalk. The banana authority Simmons believes potassium (third number on the fertilizer bag) may be more important than anything else in contributing to good fruiting of bananas. Some University of Florida agriculturists think increasing the nitrogen level gives significant results.

But, as a starter, try raising the levels of potash (potassium) and magnesium by adding sulfate of potash and Epsom salts. If you have access to quantities of seaweed, pile it on. It's rich in potassium and magnesium.

Q. Does cutting off the flower add size to the fruit on a banana's stalk?

A. Yes. Chop it off after all of the "true hands" (not the tiny inedible ones) have formed. The cut may be made about six inches below the last hand.

Q. How do you know when bananas are ready to pick? I'm not a very good guesser!

A. It takes some experience to predict accurately when the time is ripe. Generally, they are ready when all the angles on the fruit have rounded out and the bananas begin to crowd each other. That's about 120 days from the time the bloom appears.

Hang them in a cool, shady place such as a carport and within two weeks they will be ready to eat. Start cutting the hands from the top of the bunch. Tree-ripened bananas are not best for eating; they usually split.

Q. I would like to plant some papayas and would like to know how to plan a spray program.

A. You will get good results by spraying with Sevin and sulfur or chlordane and sulfur about the time the fruit sets. This helps prevent damage from the papaya fruit fly, whitefly, and mildew. Keep a close watch for hornworms on the leaves. They are easily removed by hand-picking.

Q. My neighbor tells me it is necessary to have a male and female papaya tree to get fruit. Is this correct?

A. There are some strains of papaya — seedlings, not hybrids — that have male and female forms. However, at least one male plant should be in your garden area to pollinate the "female" plants. There are also bisexual seedling strains that produce some perfect flowered and female plants. With these, it is not necessary that a male plant be provided since pollen will be produced for themselves and the female plants.

Q. I have a papaya tree with buds and blooms that fall off instead of forming fruit. Would you tell me what the plant needs?

A. First of all, there are some varieties of papaya which seemingly cannot hold their fruit while the plants are young. They shed blossoms for a period of months and then begin growing fruit. The varieties grown in South Florida have become so mixed that it is impossible to predict this happening.

Aside from that, papayas often do not receive as much fertilizer and water as they need to do their best. The home gardener can fertilize plants every two weeks. Watering should be done twice a week unless the soil drains slowly. Don't flood them. A good mulch also is helpful.

Another possibility is lack of pollination.

Q. I received some papaya seeds, planted them, and now I understand they could be male, female, or bisexual, a problem I didn't anticipate. How do I tell who's what?

A. Papaya genders can be determined only when flowers are produced. Male plants have long hanging panicles on which the flowers are produced in clusters. These plants ordinarily do not bear fruit.

Female and bisexual plants have flowers clustered at the base of the leaf next to the stem. On the female plants the flowers lack stamens so they must receive pollen from other plants in order to produce fruit.

Bisexual plants have complete blossoms with both a pistil and 10 stamens, but there is frequently a tendency for these plants to revert to either male or female.

Bisexual plants that are more male than female may have flowers with pistils only during the late winter and spring months so the plants will be barren much of the year.

Q. Why should my papayas come in odd shapes and often with creases? My friends don't have this problem.

A. Bisexual papayas that are more female than male will have blooms with less than 10 stamens. When two or more stamens are missing, the fruit will be more or less misshapen and may have a crease on one side.

Rotten spots frequently form in these creases before the fruit reaches maturity.

Q. What can I do about worms in my papayas? All my fruit is infested with them, but I notice that the papayas from the market don't have them. Nothing I've tried has helped.

A. These worms are the larvae of the papaya fruit fly, sometimes called a wasp. They are not found everywhere papayas are grown, and neither are they much trouble to commercial papaya growers here, since they only seem to infest trees on the edges of the groves.

Where trees are found in small numbers, however, they can be a considerable nuisance. If you don't mind using a highly toxic spray like chlordane you can achieve some control with this chemical. Otherwise, the safest solution is to paper-bag the fruit. (See also p. 120.)

It's too late to attempt control after the female fruit fly has deposited her eggs, so bagging should begin when the fruit is small, shortly after the flower parts have fallen. Each fruit should be enclosed by a three to five pound-size bag tied around the fruit stem. Newspaper (a half-sheet) may also be rolled to enclose the fruit, then tied around the fruit stem and the free end.

In this process, the fruit must be handled carefully to avoid damaging, and the bags should be checked at weekly intervals to be sure the young fruit is covered.

When the fruit is about the size of a big grapefruit, the bag may be removed. The fly may still be observed trying to lay her eggs, but the ovipositor cannot reach the cavity at this point.

Q. I've had bad luck with papayas and I don't seem to be doing any one thing wrong, I seem to be doing everything wrong. I correct one condition and then someone tells me that the fertilizer is not the best. I try another kind, then my nurseryman tells me its feet are probably too wet. Give me some guidelines and I'll start all over.

A. Four hours in standing water and it's usually all over for papayas. So plant them high and dry; make your own miniature "hill" if necessary, about 12 to 18 inches higher than the rest of your yard. Mulch will help retain necessary moisture and keep down root-knot nematodes.

Papayas need twice as much plant food as most other plants, and any organic material should be supplemented by a 4-10-12 formula or fish scraps and chicken manure, plus superphosphate. You can get more information from your county agent, or you can write to the Hawaii Agriculture Experiment Station, University of Hawaii, Honolulu, for their special tells-all bulletin.

Q. We have a five-year-old loquat tree that blossoms profusely but doesn't produce fruit. Can you tell us why?

A. If you have a seedling tree, it would be normal for it to bloom but not produce fruit, although not all seedling loquats are fruitless.

If it isn't, make sure you have it on the right fertilizer program. Try a complete mix of 6-6-6 and use nutritional sprays.

Q. I've got multi-colored spots on my loquat. They start out reddish brown, then as they get bigger, they develop a purplish margin with sort of a yellow-green circle around it. Next the center gets an ashy brown and blisters. The ultimate result is that my tree isn't going to have a leaf left on it if I don't do something. What?

A. Your tree is a victim of entomosporium leaf spot, and the first step you should take is to remove and destroy all infected leaves that have fallen to the ground. The fungus spores that cause the spot thrive in old, infected leaves. You might also check other nearby plants for the spot. A fungicide like a fixed copper or Maneb, Zineb, or Ferbam should be applied.

Q. At what time of year does a loquat tree bear fruit? Ours blossoms right along but does not form fruit. What fertilizer should we use?

A. The fruit is ready in late February or early March. Lack of production could indicate your tree is a seedling. Otherwise use a complete 6-6-6 mix with nutritional sprays.

Q. I bought a grafted carambola tree, now about six feet tall. What can I do to make it bear fruit? I've sprayed it with a nutritional spray containing zinc, and I fertilize and water it regularly. It bloomed about a month ago, but the flowers dropped off after blooming. What am I doing wrong?

A. Whether your tree was container-grown or field-grown, give it a year to establish itself. Your tree has been in the ground only a few months. It needs time to grow a strong root system and gain some size. It wouldn't be at all unusual for it to shed the bloom now without setting fruit.

You can push the tree by fertilizing monthly the first year. The application should be light. Continue to water when needed. You might use the nutritional spray three or four times this first year.

Once the tree is established, you can cut back to three or four fertilizer applications a year and one or two applications of nutritional spray.

Use a fertilizer with equal amounts of the major elements — 6-6-6 is fine, but it should contain at least three percent magnesium.

Keep in mind, though, that if the tree was not grafted, seedling carambolas can have pollination problems.

Q. My carambolas have some brown blisters on them which seem to be caused by a lot of little green-blue flies buzzing around the fruit. I've put plastic bags over the carambolas, but the flies continue to swarm and the fruit doesn't look any better. How can I get rid of these pests?

A. Those irridescent flies are no threat to your carambola; however, the plastic bags you are using definitely are.

Plastic bags will increase the humidity around the fruit to the point where anthracnose is very easily formed or aggravated. The "brown blister" you describe may be anthracnose. Try taking the bags off, spraying with a neutral copper solution, and waiting to see if the condition clears up. Carambolas aren't subject to damage by flies, but sometimes a puncture or otherwise damaged fruit may attract them.

Q. My guava bears very good fruit and plenty of them. The problem is that little white worms appear in the ripe fruit even while it is still on the tree. I have sprayed the green fruit with Malathion every two weeks, but the ripe fruit is still wormy and inedible. Can you help?

A. Only to the extent of identifying the little white worms — larvae of the Caribbean fruit fly. The guava is perhaps this pest's favorite host but it

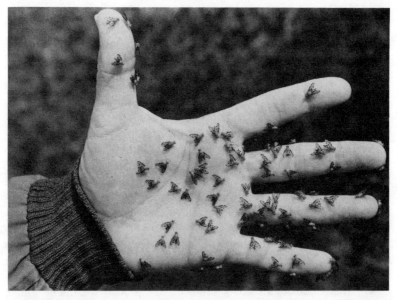

The Caribbean fruit fly is a major pest of several soft-skinned tropical fruit trees in Florida and occasionally will lay its larvae in thicker-skinned fruits like the citrus. Various control programs have been sponsored by the state and the University of Florida, but financing massive control has not yet been possible.

also ruins Ceylon peaches, loquats, Surinam cherries, sapodillas, eggfruit, kumquats, calamondins, rose apples, and occasionally other fruits.

The fly lays eggs beneath the surface of the fruit skin and they hatch into the worms.

Experimentally,DeFend or Cygon (dimethoate), Dylox, and Formothion (Sandoz) have been used with good results. Dylox, for example, has been mixed at one teaspoon per gallon of wettable spray in a trombone sprayer and used once a week, starting when the first fruit showed color. In the rainy season, it has been used successfully once every five days. Spraying was halted a week before eating the fruit.

But these chemicals are not yet licensed for use on fruit trees and so no one can recommend their use. They are unlikely to be licensed in the foreseeable future because of the expense of the testing needed to prove them safe on fruit to be eaten.

Weekly spraying with Malathion, a licensed material for fruit trees, although even it may not have general approval, could control the problem to some extent if begun early enough.

Q. Is it possible to grow peaches in South Florida? If so, what is the best variety?
A. About the only variety that does well in South Florida is the Red Ceylon. However, peaches are a prime host to the Caribbean fruit fly and until this "spoiler" is brought under control, it will be very difficult for you to produce an edible crop.

Q. Help! My poor fig tree was doing so well but now the leaves have turned brown and fallen. I did nothing at all to it, but we live on the bay. Could this be a factor? I just don't know what to do.
A. There are several kinds of figs, and most have diseases special to each. There is a strong possibility your leaf drop is the result of fig rust fungus. Control for this is application of a neutral copper spray applied monthly in early summer. Figs share at least one thing in common. They are not tolerant of salt conditions which you have on the edge of the bay.

Q. I've had some trouble with nematodes and I understand this makes me ineligible for fig growing. Isn't there something I can do to the soil? If not, I'll have to be content watching (and eating!) my neighbor's.
A. Where nematodes are a problem, try ground treatment with a nematocide and planting your fig next to a building. This way the roots can penetrate the soil under the structure, where nematodes are fewer. Nematode damage can also be lessened by heavy organic mulching and by a

dense planting of marigolds around the fig, a measure supported by organic gardeners but doubted by others unless a thick planting is maintained.

Q. What kind of figs grow best here?
A. Figs have a nematode problem in South Florida and are best grown grafted on the rootstock of one of four other species of ficus: *F. coculifolia* from Madagascar, *F. glomorata (F. racemosa)* from India, *F. gnaphalocarpa* from Angola and West Africa and *F. palmata. F. glomorata* and *F. gnaphalocarpa* are both seen growing successfully in South Florida.

As to a good variety for budwood, the large green lemon fig is probably the best. 'Brown Turkey' is fine, too, and others used are 'Celeste' or sugar fig and 'Brunswick.'

Q. Please give me some information on the care and feeding of a lychee tree.
A. An established tree can be fertilized three times a year — say, in March, in June or July, and in late October or early November. The last application perhaps should be lighter than the others. Use a nutritional spray in January.

An application of chelated iron can be made once a year. If foliage tends to be yellow, more iron can be used. A good layer of mulch over the root zone will help. Don't use thick layers of green mulch. Keep the mulch a foot away from the trunk. Remember to water during dry periods. A thorough irrigation every two or three weeks may be enough.

Q. I'd like to air layer my lychee to be able to give a couple of trees to my brother. My questions are: (1) Can you use lateral branches? (2) Do air layered plants bear fruit sooner? and (3) Are they as rugged?
A. (1) If the tree is in good condition, any branch with its terminal leaves in sun will air layer readily, (2) there's a possibility, but there's no scientific proof, and (3) there's no reason why they shouldn't be, if planted in a properly prepared site. Almost all of South Florida's lychee trees have been grown from air layers.

Q. Last year I had a wonderful crop of lychees. This year, nothing — not even much in the way of flowers. Could this be a result of the weather?
A. No. Lychees — the 'Brewster' variety, in particular — are strongly alternate bearers, which means good eating every other year. Lychees have been known to bear heavily two years in a row, however, breaking the rules. This unusual turn of affairs doesn't necessarily mean that lychee

growers will have two fruitless years. Little or no bloom is standard behavior the year following a good harvest.

Q. My Barbados cherry tree was purchased about a year ago from a reliable nursery. It is about seven feet tall, the foliage is full and healthy looking, and there have been several blossomings. Yet the fruit rarely reaches maturity. Either it drops off immediately after the blossoming or later on.

The tree has been fertilized with a special fruit fertilizer about every three months and I water once a week with about one inch of water, except lately during the drought when I water somewhat more frequently.

A. Here are some possibilities:

The tree has been in the ground only a year. It still may not be established well enough to produce a crop of fruit. After moving from nursery to dooryard, many plants spend a period of time in putting out an extensive root system and increasing in size above the ground as well.

Many fruit trees don't seem to do well in producing a crop if they are pampered too much. Continuing heavy watering and heavy fertilization may keep the tree increasing in size without setting much fruit. You might experiment a little. The tree, of course, should have enough water to keep it from wilting, and enough fertilizer to keep it in healthy condition.

If you have not used a nutritional spray, this might help. Such a spray provides mineral elements the plant can use. Spray at a time when most new leaves have matured.

An organic mulch also is beneficial. It adds organic material to the soil, plus its other good effects.

Q. My pomegranate looks healthy and produces plenty of fruit, but the color of the fruit is dull and the flavor can't compare with the fruit I buy in the market. What's wrong?

A. Pomegranates like long, dry summers. In our humid climate they do not color well and their sugar content is usually low. Often the fruit splits open. Pomegranates are more satisfactory as ornamentals and the double-flowered types are especially attractive.

Q. I have pomegranate plants, but they don't seem to get green, blossom, or produce fruit. I fertilize well, but no results.

A. Continue your fertilizer program, but also make an application of chelated iron on the soil around the plants. This material is available from garden supply stores and must be applied to a well soaked soil to be effective.

These plants often will drop their leaves or most of them during the winter, but new leaf growth and bloom usually are seen in the spring.

The leaves you sent had dark spots caused by fungus. This is common on pomegranate plants, and you can try for control by spraying with neutral copper. A mulch over the root zone of these plants probably would be of some benefit.

Q. My sapodilla's leaves began turning yellow and my nurseryman recommended spraying it with chelated iron. Some of the leaves regained their former healthy green but others didn't. Now what do I do?

A. Get out the spray gun and finish the job. You evidently missed some. Iron can't be transferred from one leaf to another, consequently only leaves receiving soluble iron turn green. Likewise, if a single leaf is not thoroughly covered by the spray, it will turn green only on the spots covered by the spray.

Foliar applications of iron generally do nothing more than freckle the leaves. Apply it to the soil.

Q. I have been told that macadamia trees can take as long as 20 years to bear but I still want one. How old a tree should I buy or is there a way to make these trees bear in a shorter period? How long does it take for the nut to develop?

A. Macadamias take from 10 to 12 years to fruit from seed but you can cut the fruiting time to three to six years by buying a good macotted (air-layered) variety, a grafted tree or one grown from terminal cuttings. The fruiting season is from November to March, and the nuts mature in about six or seven months after flowering.

Q. I look forward to making jelly and salads from the carissa I just bought. What's the secret of raising good fruit?

A. Natal plums like light, well-drained land and are very salt-tolerant, making them one of the state's best seaside plants. They should be mulched, and two or three applications of fertilizer are recommended each year. Keep grass out of the root zone. Full sun encourages enthusiastic fruiting.

Q. I have a sea grape next to my patio that is about six years old and so far I haven't seen any fruit on it. Now I hear that I may be stuck with a male sea grape. Is it true that they come in sexes and, if so, how do I tell what mine is?

A. The rumors you've heard are true but it is impossible to distinguish males from females by foliage or growth characteristics before the telltale flowers appear. The trees begin to flower and fruit anywhere from six to eight years from seed.

Give your tree another year or so, then if it doesn't fruit you may start investigating two possibilities: (1) it's a male, or (2) it's a female but there is no male sea grape in the neighborhood to facilitate pollination.

In the meantime, the best course of action would be to plant some more good-size trees, taking the chance that one of them may turn out to be female, in case you're the owner of a male.

Q. How far in north Florida can you grow jaboticabas?
A. Many exotic fruit trees flourish along the Indian River as far north as Merritt Island, side by side with coconut palms and Australian pines. Temperatures are, of course, warmer along the coastal areas than a few miles inland where the same fruit wouldn't be able to take the winter weather.

Even Indian River growers, however, have to steel themselves against periodic losses from unusually severe weather such as the freeze of '61-'62 when Indian River homeowners were parted from their exotics.

Q. I love all of the "standard" fruit — citrus, mangos, avocados, etc. — I have grown over the past years but now I'm ready for something new. Macadamias and cashews are not for me — I want something I can eat right off the tree!
A. One delicacy that fits your requirements is the "African Pride" atemoya, a cross between the cherimoya and the sugar apple. The cherimoya is the earliest recorded fruit of the New World and was used as a model for prehistoric Peruvian vases. The sugar apple is native to Central America and the West Indies and has been grown commercially on the Florida Keys.

Both of these belong to the annona family, as does the custard apple, which is recommended as root-stock for the "African Pride."

Q. I understand that the akee is poisonous but it's such a spectacular tree I have to own one. Can you give me a few details about the fruit?
A. The akee's Latin name, *Blighia sapida*, honors Captain Bligh of "Bounty" fame. The big, showy red fruit comes in three parts and when these pop open on their own, the aril that holds the shiny black seed is edible.

Before they open, and when they're overripe, they are poisonous. In between, they're delicious.

Native to West Africa and common in the West Indies, they are a favorite

Jamaican breakfast dish. Akees need heavy mulching and occasional applications of a 6-6-6 fertilizer. Propagation is by seeds and shield budding.

Q. A friend in Mexico who knows I like growing fruit trees has offered to bring back a black sapote if I'd like to experiment with it. Can they be grown here and, if so, how successfully?

A. The black sapote, a persimmon relative, seems to be growing in popularity in South Florida. It thrives in alkaline soil, responds readily to care and fertilizers, and is comparatively free from disease and insects.

The unusual color of the four-inch fruit — a black or dark brown flesh — has earned it the name of the chocolate-pudding fruit. It's tasty and high in ascorbic acid, calcium, and phosphorus. It also contains thiamine, riboflavin, carotene, iron, and niacin. The sapote's dark green leaves and symmetrical shape make it an ornamental addition to any garden. Protect it from wind damage during its early years, and give it a nutritional spray twice during the peak growing season to discourage a tendency toward magnesium deficiency.

Q. I planted four pineapple tops, and each has taken nicely. One plant became quite huge and the pineapple that came out is almost the size one buys. Now there is a stalk growing right in back of the pineapple. Will this also bear fruit, and do I take out the plant and put it elsewhere?

A. The stalk you mention probably is a shoot which can be left where it is or removed and planted to produce a new plant. Each pineapple plant bears only once, and new plants are grown from suckers that appear at the base of the plant, shoots which grow at the base of the fruit, and, as you did, from the tops cut from mature fruit.

If this stalk is from the original plant it will bear — but subsequent sprouts will tend to fall over.

Q. If I buy a pineapple in a supermarket and take off the top and plant it, what special care will it need? How long will it take to produce fruit?

A. After removing the top, let it dry for a few days. Plant it in a container of good soil and fertilize with a liquid fertilizer. After the plant is well rooted, it can be set in a bed. A 6-6-6 fertilizer can be applied on the soil, and you can continue to apply liquid fertilizer to the leaves. With good care, the plant may bear fruit in a year to a year and a half.

Q. I have been given a pineapple plant in a gallon can and would like to transplant it in the yard. What is the best way?

A. You will need a well drained location for a pineapple plant, and the soil should be improved by adding peat moss or humus. The drainage is important because these plants can't stand to have their roots in water for any length of time.

Fertilize the soil before you plant with composted manure. A 6-6-6 fertilizer can be added to the soil three or four times a year.

The plant also will benefit from liquid fertilizer sprayed on the leaves each month or so. If you think your soil may be infested with nematodes, you should apply a nematocide to the soil two weeks before planting. Established plants won't need much water. Twice a month should be enough watering in the dry season.

Q. I've got a variegated pineapple, but I'm certain it's not as brilliant as it was when I first got it. I've put it in my shade house. Also, can it be propagated from seed?

A. This plant could very well revert to green, become streaked with yellow, and lose its bright pink and red variegation. The best balance is said to be about 40 per cent variegation and 60 per cent green so there will be enough chlorophyll to make sugar for the plant. Move your plant gradually into more sun and keep an eye on its response.

Variegated pineapples can not be propagated by seed. Use the offsets that form at the base of the fruit but let them get a little beyond babyhood before you remove them from the plant.

Q. Birds are getting to my fruit before I can. They're terrible this year. What can I do?

A. The birds' big appetites may be due to a smaller crop of fruit in your neighborhood than usual, but there's nothing that can be done to discourage them, short of bagging each fruit you want reserved for you. There's no spray, or bird alarm that will take care of the problem. Dangling pie plates or metal strips don't impress birds for long. The best approach is to plant enough fruit trees for both of you.

Q. Have you any idea how to take out fruit stains? I'm an impulsive gardener and don't always take the time to change when I should.

A. Avoid using soap on fruit stains. It tends to set them. Veteran stain removers recommend boiling water poured over the area at once. If the fabric would be damaged by boiling water (don't use this on wool, for example), sponge it with hot water, then rub it between the hands with a little glycerine. Let it stand for a few hours, apply a few drops of vinegar, then rinse off thoroughly in water.

7. Vegetables

*For special index to this chapter,
see back of book.*

In the topsy-turvy world of vegetable growing in the tropics and sub-tropics, backyard farmers need help even before they buy their seed. Seed planting months in South Florida, for example, are usually October and February, with the spring and summer months reserved for a reduced menu of vegetables that can take both the temperature and rainy weather. (Purchased seedling plants can be set out a little later than seeds are planted and still produce before running into weather problems.)

Regardless of the season, there are varieties easier to grow in the tropics than others more familiar to northern gardeners. Destructive soilborne organisms and nematodes are more common, and soil is often treated for these before planting. Weekly spraying with Sevin and a fungicide are good practices. Raised beds filled with topsoil and humus are common.

The ins and outs of growing vegetables in your area are easily checked with your county agricultural agent who's eager to help small growers as well as large. He can save you time, money, and effort with a phone call.

Q. What is the best way to prepare your soil before planting vegetables?

A. If you are on a fairly deep, sandy soil, you should mix in peat moss (according to instructions on the bag), humus, or compost to help retain fertilizer and water.

If you are on rock soil, a raised bed makes the whole project easier. Use cement blocks, railroad ties, a pressure-treated wood frame bed, or even coral rock to mark off its boundaries. Buy topsoil from your nurseryman and have it emptied into the "frame." This way you can grow root crops and your soil will hold water and fertilizer better, especially if peat moss or compost is added.

Marl soils need little done to them outside of fertilizing.

Unless you're a hopeful organic gardener, fumigating with a nematode-killer is provident.

Q. How should you fertilize a vegetable plot? I get conflicting advice.

A. There's not just one way, but a popular approach is to buy 100 pounds of 6-6-6 (30 percent organic) for a 20-by-30 plot. It will meet all its nutritional needs for a season if you put down half when you plant, then the rest in two- to three-week intervals after the seeds sprout.

Another approach is to use Milorganite when you prepare the bed, then fertilize lightly between rows every two weeks with a 6-6-6.

Some backyard farmers use organic fertilizers heavy on nitrogen to bring their crops to adolescence, switching to a 4-8-8 with those that produce fruit rather than just leaves.

Vegetable gardening in Florida can be fun or acute frustration, depending on the gardener's degree of determination to "follow the rules." Selection of varieties recommended for this area and weekly preventative spraying with both insecticides and fungicides are factors in the most successful vegetable plots.

Q. I'm interested in avoiding disease problems with my vegetables. Bugs I can see and spray for, but my neighbor says there are fungi, bacteria, and viruses around that just make growing vegetables too much of a bother. I can't quite accept this, with the big, profitable vegetable industry we have in South Florida.

A. A little forethought and a little free advice from experts is your ounce of prevention. When vegetables give up the ghost for no apparent reason, it's often due to diseases caused by unfavorable environmental conditions — too much or too little soil, moisture or sunlight, a nutritional deficiency or excess, or temperatures too high or too low for the particular vegetable. Find out the light, soil, food, and moisture requirements beforehand for each vegetable you plan to raise (they vary). The surest way to get correct information is by calling your County Cooperative Extension Service (listed under "U.S. Department of Agriculture" in your phone book).

Some of the fungicides used to control diseases on vegetables caused by viruses are Captan, Chloranil, copper, Ferbam, Maneb, PCNB, sulphur, Thiram, and Zineb. Your county agent can diagnose a virus problem.

To guard against nematode damage, many gardeners treat their soil with Nemagon before planting. Another precaution is to avoid using weed-killers around vegetable plots. And a third is to buy bedding plants, if they are available, of disease-resistant varieties developed for South Florida. This especially applies with tomatoes.

Of course, farmers have recourse to more potent chemical controls for problems than homeowners, so don't worry about not growing absolutely flawless vegetables.

Q. In a garden book I read, I noticed the terms "top-dressing" and "side-dressing." What's the difference between applying fertilizer to the "top" or to the "side"?

A. These terms virtually have the same meaning — an additional light application of fertilizer made later in the growing season to supplement one or more of the initial basic fertilizers previously applied.

Since the feeder roots of many vegetables extend some distance out from the plant — unlike most root crops — a "side dressing" between rows is made.

Q. Does the kind of fertilizer used on a garden affect the taste of vegetables?

A. There isn't any evidence to show that fertilizer changes the taste of vegetables in any way. Taste is determined by other environmental factors and the vegetable varieties themselves.

Q. I used a commercial material for nematodes before planting tomatoes, cucumbers, and peppers. The cukes turned yellow. A friend, who planted the same as I did, had tomatoes which prospered. Why?

A. Sounds like the soil didn't get proper aeration after you used the nematode chemical. Yellow-appearing vegetables often indicate fumigants still retained in the soil. Perhaps you planted too soon after using the nematocide.

Q. What vegetables should be planted directly in the garden as seed rather than grown in flats?

A. If it's slower to harvest, generally speaking, like celery, cabbage, tomatoes, peppers, and eggplant, it can be grown in flats, then transplanted. It's almost impossible to transplant successfully members of the cucurbit family like cucumbers and squash, and beans, turnips, and corn. Carrots, radishes, and kale are medium-easy to transplant.

Q. What vegetable varieties are recommended for South Florida? How about 'Beefsteak' and 'Better Boy' tomatoes?

A. South Florida grows gorgeous vegetables, but you need every advantage you can get since it grows great fungi, bacteria, and various viruses as well.

Of course, some people have had good experiences with varieties that haven't been bred resistant to Florida problems, such as the tomatoes you mention, but these stronger vegetables cause fewer problems:

BUSH BEANS. Extender, Contender, Harvester, Wade, Cherokee (wax).

CABBAGES. Copenhagen Market, Marion Market, Badger Market, Glory of Enkhulzen, Red Acre, Chieftan Savoy.

CARROTS. Michihli, Wong Bok.

COLLARDS. Georgia, Vates.

CUCUMBERS. Poinsett, Ashley, Wisconsin, SMR 18, Dixie.

EGGPLANTS. Florida Market.

LETTUCE. (Crisp) Premier, Great Lakes types; (Butterhead) Bibb, Matchless, Sweetheart; (Leaf) Prize Head, Ruby, Salad Bowl; (Romaine) Parris Island Cos, Dark-Green Cos.

MUSTARD. Southern Giant Curled, Florida Broad Leaf.

ONIONS. (Green) White Portugal or white types. Plant September through March (Bulbing). Do not plant after November: Excel, Texas Grano, Granex, White Granex, Tropicana Red.

PARSLEY. Moss Curled, Perfection.

PEPPERS. Plant through October, then not again until January. Varieties are California Wonder, Yolo Wonder, World Beater (Sweet), Hungarian Wax, Anaheim Chili (Hot).

RADISH. Cherry Belle, Comet, Early Scarlet Globe, White Icicle, Sparkler.

SPINACH. Virginia Savoy, Dixie Market, Hybrid 7.

SQUASH. Plant through October and not again until January and February. Varieties: Early Summer Crookneck, Cocozelle, Zucchini, Patty Pan.

STRAWBERRIES. Plant October and November. Florida 90, Daybreak, Torrey.

TOMATOES. Manalucie, Homestead-24, Indian River, Floradel, Tropired, Atkinson, Large Cherry, Roma, Sunray (yellow).

Q. I've had some bad experiences with garden peas. Is there anything else that doesn't do well in South Florida?

A. You can grow English or garden peas at the tip of Florida if they are not planted before the first of November and not later than the first of February. Artichokes and asparagus are seldom grown well here. Rutabaga is not spectacular in this area. The same can be said for Brussels sprouts. Head lettuce is tricky in a mild winter but grows well.

Vegetables that do better in the summer than winter are squash, Southern peas, okra, corn, melons, and sweet potatoes. Cucumbers sometimes will make it through hot weather, too.

Q. How can you prevent birds from pecking at strawberries? I have the plants growing through plastic which is excellent for weed control, but the birds are posing a very serious problem.

A. If you have a home garden, stretch a piece of transparent plastic over the plants. This will permit enough sunlight to get through but will prevent the birds from getting at the berries.

Q. I have tried the strawberry barrel, and the results have been very satisfactory so far, but recently I have noticed an apparent disease. Little red spots appear on the leaves and they eventually wilt. What is the treatment?

A. A leaf-spot disease may be the cause of the problem. For control you could use, weekly, a spray or dust containing Captan or Benomyl. Check your garden supply store and follow all directions and warnings.

Q. Several years ago I read that by proper selection of varieties, one could raise corn here all year long. I wrote my county agent and received

a circular about corn, but it does not provide the varieties and the planting schedule.

A. The question is not so much one of variety but of insect control. Plant a good variety, such as Seneca Chief, in the summer and it will grow, but the insect problems are much greater than during the winter.

In the home garden, the agricultural office suggests spraying corn every three to five days with Sevin. It is important that the spray material reaches the silks. Hopefully, your corn will then be free of worms.

Q. All the plants in my vegetable garden did nicely last year but soon after flowering and starting to bear crops they died. I treated the soil with cow manure and fertilizer. Did I do something wrong? Could there be a disease in the soil?

A. Probably two things. First, you apparently didn't fumigate the soil. Second, you do not mention any kind of spray program, and vegetables like tomatoes, cucumbers, and other cucurbits can decline very rapidly when they aren't being sprayed regularly with insecticide-fungicide material.

Q. I planted some sweet potato vines, and now some of them are blooming? Is this unusual? The flowers are about two inches long and are rather pink in color. My neighbors keep asking if I'm sure I planted sweet potatoes.

A. It doesn't happen too often, but sweet potato vines do bloom occasionally in the U.S. It happens more often in the tropics where they are native.

The bloom may remind you of the morning glory. The plants are closely related. Sweet potato is *Ipomoea batatas,* and common morning glory is *Ipomoca purpurea.*

Q. Will you tell me how long it takes sweet potatoes to mature? I planted some vines in my yard and they are doing well.

A. You can expect the roots to be mature four to five months after planting. A weevil can cause problems with this plant. The county agent's office suggests Sevin spray to control this pest.

Q. Why do my cucumber plants have loads of blossoms and no cucumbers? I have tried for a number of years, and no cucumbers.

A. Suspect that the blossoms are not being pollinated. If you don't see any bees visiting the blooms, particularly in the morning, this likely is the answer.

You can try hand pollinating a few. First, you have to know the difference between the female and the male blossoms. The female will have what appears to be a miniature cucumber at the base of the flower. The male bloom has no such formation.

Usually, there are noticably more male flowers, and a vine may have several of these before a female bloom appears.

To pollinate, pick a male bloom early in the morning — say before 9 a.m. Hold it over the female bloom and tap it to dislodge the pollen. That's all there is to it. Some people use a brush, too.

In a few days you should notice that the miniature cucumber has started to enlarge. You can use the same process if you attempt to grow squash, pumpkins, watermelons, and similar crops and the bees aren't cooperative.

Q. I have been here four years and cannot grow lettuce, since little white worms set in and eat it before the lettuce develops. What can I do?

A. Your problems are coming from a caterpillar-type worm. Spray the lettuce patch with Sevin.

Q. I have planted watermelon in my backyard, they grow, but won't fruit. The little melons fall off. Why? The yard is sand, and I have seen them grow here in sand. I fertilize and give them water.

A. When melon and squash vines bloom well but don't set fruit, the usual answer is lack of pollination. Where there are enough bees and other pollinating insects, they do the job, carrying pollen from the male bloom to the female blossom.

If insects don't do the work, the gardener can hand-pollinate the blooms. You can tell the difference between male and female blossoms by checking the base end of the flower. The female will have an enlarged base shaped like a very small melon. This is absent in the male bloom.

A vine usually produces more male than female blooms, and several males often are found before a female flower develops.

When you find both male and female blooms on the vine, pick a male flower and hold it over a female blossom, tapping the male gently to discharge pollen. This should be done fairly early in the day, before 9 a.m.

Q. I have had a great deal of trouble trying to grow zucchini squash. The first bloom that comes out is perfect but the bloom that grows with the squash rots and then causes the squash to rot, I assume.

A. Faulty pollination is a frequent cause of problems with squash, pumpkins, and melons. Pollen must be transferred from the male blossom to the female, and if there are no bees or other pollinating insects to do the job, the gardener will have to hand-pollinate the plants.

The first blooms to appear usually are male, and the male blooms usually outnumber the female. The female bloom is the one with a miniature squash at the base.

A male bloom can be picked, held over the female bloom and tapped gently to transfer the pollen. This should be done in early morning, before 9 a.m.

Q. I'm having a problem locating the seeds for tomato plants developed especially for Florida. Any suggestions?

A. Your best bet is to buy the plants at your nursery. If you want to start from scratch, however, the "ordinary" standard varieties of seed will serve your purpose on a small scale. They're just not recommended for larger plantings because they require more maintenance in Florida. Many of the catalog seed companies carry 'Homestead,' 'Floradel,' and 'Tropired,' the most disease-resistant.

Q. I'd like my tomato problems diagnosed. When I bought my plants they were in bloom and healthy looking. They went downhill from the time I planted them. First holes appeared in the leaves, then the branches died back. I put on a tomato dust but couldn't see a change. I love tomatoes and would like to save what's left of these plants. Is there any hope?

A. Your experience shows why it pays to consult your County Cooperative Extension Service agents before wading into unfamiliar territory. First try to locate one of the varieties that are adapted to Florida conditions. These have multiple disease resistant qualities necessary for best results here.

Next, a plant that was left in a pot long enough for it to develop blossoms was probably rootbound and would never develop into a large, vigorous vine. In the future, try purchasing plants that haven't reached this stage.

Before you purchase a control material, you should know what you are trying to control. Best way is to send a sample to your county agricultural agent and describe the problem in detail. Every nurseryman can't be expected to know all about diseases and pests on both ornamentals and vegetables.

Most tomato dusts or spray products sold in South Florida contain material like Maneb to control diseases such as early and late blight and a general purpose insecticide like Carbaryl. However, yours may have contained Captan instead of Maneb, and if so, and your problem was severe late blight, the Captan couldn't have coped with it.

Bacteria spot is often responsible for branches or leaves dying, and it can be controlled with a copper fungicide. The holes in your leaves aren't related to this problem; a general purpose insecticide can take care of the muncher who made them.

One problem your tomatoes don't seem to have — but which is common — is leaf miners. Their presence is identifiable by white trails across the leaf surface. Malathion will stop them.

For better success next time, keep in mind the improved varieties and young started plants.

Q. My problem is the sudden death of my tomato plants. At first, they are very green and luxuriant. Suddenly they look rather wilted. The leaves turn brown, one section begins to droop, and they break off. Could nematodes cause this?

A. Your description of the problem doesn't fit nematode damage. Nematodes cause the plants to be stunted and to do poorly, but it usually is a gradual process.

Your problems sounds more like one of the wilts which hit tomato plants. There are soil-borne diseases which would not be affected by the spraying you do to control insects and fungus.

Easy to grow in hot summer weather is the cherry tomato, a sweet, succulent treat that is especially favored by dieting gardeners who need to nibble on something good. Children also enjoy growing this bright little bush.

Some tomatoes such as the 'Homestead' are better able to resist these diseases than others.

If you want to grow the tomatoes more common farther north you might consider sterilizing the planting bed, or planting the tomatoes in large containers of sterile soil.

Q. I have treated cans of soil with Nemagon for nematodes. After the tomato plants are in and have grown and are bearing, will any of the chemicals and poison be absorbed into the tomatoes themselves? Will they be safe to eat? Also if onions are planted in same treated soil, will the grown onions be safe to eat?

A. The county agents say the tomatoes would be safe. If you plan to grow the onions to large size for use as dried onions, you can plant in treated soil, but the treated soil would not be recommended for growing green onions.

Nematode killers are valuable aids in preventing root-knot and other problems caused by the microscopic soil pests, but all directions and warnings on the label should be followed exactly. The labels are prepared to assure safe use.

Q. I recently sterilized the soil in a patch of ground with some VC-13, leftover from my lawn. Then I planted tomatoes, I have now been told that the fruit will not be fit to eat, as the tomatoes will absorb the poison from the ground. Is this true?

A. You needn't worry that your tomatoes will be poisonous. VC-13 is not a systemic insecticide, and will not be taken up by the plant. It also is not a very effective nematode killer, by the way.

Any gardener who uses a pesticide should follow all directions and warnings on the label. Some will say the product should not be used on food crops. Others will say which food crops they may be used on.

In fact, it's a good idea to read the label twice. And be sure you understand what you read. A survey in the Midwest recently turned up the information that many people didn't know what the words on the labels mean.

Actually, what you did will not "sterilize" the soil, but it should reduce some soil pests. A planting bed is sterilized with a chemical which forms a gas under a plastic covering.

Most gardeners won't go to all that trouble. Many go ahead and harvest a reasonably good crop from infested soil. They do this by providing more fertilizer than the plants normally would need.

Q. I've been told that a "railroad" worm is a problem with my

tomato leaves. Can you tell us how to eliminate this pest. Could it be from cow manure I used two years ago?

A. Your problem is the result of a leaf miner. Spray with Malathion. The cow manure did not cause the problem. The miner comes from gnat flies.

Q. We have used Malathion against an insect that bores into unripe tomatoes. However, we are now afraid to eat the tomatoes since we understand that Malathion is very closely related to the deadly poison, Parathion. Since you quite often recommend Malathion, could you please tell us if it is safe to eat the tomatoes. Is it harmful for our dogs to snoop under the sprayed plants?

A. Improper use of any insecticide can be harmful. Most Malathion preparations are manufactured under safe tolerances. It is all right to use it on the tomatoes, but each tomato should be thoroughly washed before it goes to the dinner table. Wash all fruits and vegetables, whether grown in the home garden or commercially, and this includes all those bought at the fruit and grocery stores.

Q. Why do most hydroponic tomatoes seem firmer than those grown in the soil?

A. Potassium is the plant food element that has a hardening effect on tomatoes, and in the hydroponic system of controlled growth, it's easy to keep the potassium at high levels in the solution. The same can be done in soil after soil tests have been made.

Q. Is it true that organically grown vegetables are better than chemically grown? My teen-age son insists on bringing home organic vegetables for his own dinner and it's a real nuisance to have to cook separate meals.

A. Dr. B. E. Day of the University of California once presented a now-widely-quoted opinion which seems logical: "With the exception of a few parasitic plants, higher plants do not utilize organic nutrients. Plants require water, carbon dioxide, and several inorganic ions and nothing more. Water, air, and a few simple salts constitute a complete nutritional environment for green plants, and it is immaterial whether these ingredients are supplied from decaying compost or from a mine or factory."

All nutrients must become chemicals, identical in their arrangement of atoms, before they can enter a plant.

Day went on to say that for many reasons, "the obvious and sensible place to put our plant and animal wastes, in most instances, is back into the soil. There are sufficient good reasons for organic farming without giving credit or credence to the phony claims made by cultists."

Q. While up North I bought a so-called white sweet potato that is much drier than our juicy ones. I really love this particular kind so I thought I'd cut them up and plant them. To keep them from overgrowing my small place I have been cutting the leaves back. Does this take something from the potatoes? They've been in the ground for weeks but I find nothing but strings underneath.

A. Yes, restricting the size of your vines will reduce the size of your potatoes; however it's too early for you to be looking for action. Sweet potatoes take six months to start producing, and it's better if you plant your potatoes whole, not cut up. You might put up a trellis if you're running out of room and let your vines climb unclipped.

Q. I'm enclosing a leaf from my bean plant and would like its problem identified.

A. You may have a soybean infected with a soil-borne fungus called rhizoctonia, present in much of our soil. It can be controlled to some extent by using a fungicide in the furrow at planting time.

For maximum production with beans and to build up the nitrogen in your soil. It is also advisable to innoculate the seed with nitrifying bacteria, which is available through seed catalogs and at larger garden shops.

Q. I've had some problems with my first crop of tomatoes this year that I don't want to see repeated. What makes the blossoms fall off? I've given them plenty of water, fertilizer (but not too much), and sun.

A. Both dooryard and commercial tomato growers in Dade County had some trouble this year with blossom drop heavier than usual. Three main factors are responsible — high temperatures, too much rain at the time of pollination, and overfertilizing with nitrogen.

Daytime temperatures of 90 degrees at the time seed was planted and until the leaves reached an inch to an inch-and-a-quarter in length will prompt blossom drop later. Fruit set in tomatoes is very sensitive to certain temperature ranges. None of the varieties most desirable for planting in Florida will set fruit satisfactorily when nighttime air temperatures are over 70 degrees. There is practically no fruit set when night temperature is 90 degrees F. in the soil.

Also, during extended periods when the temperature during the night drops below 55-59 degrees, tomatoes will not set a good crop.

Generally, nitrogen applied in excessive amounts prior to flowering will cause the plant to continue vigorous vegetative growth rather than to set fruit. This is particularly the case when plants are growing over septic drains, in soils where a lot of peat, manure or other rich organic material has been added, or when they are fed too much liquid fertilizer.

Tomato production usually is best when only enough nitrogen is available during the first couple of months after planting to permit development of a large sturdy vine, but not enough to cause a soft type of growth.

Then, once the plant has a heavy fruit load and tomatoes begin to ripen, the nitrogen requirement is very high. If the nitrogen supply is not adequate at this stage of development, the plants will turn yellow and defoliate, leaving the fruit exposed and vulnerable to sunscale. Regulation of the nitrogen supply is an individual field and seasonal problem and must be adjusted as the season progresses.

The third factor is rain occurring at the time pollination is taking place, washing off the pollen as the blossom opens, and, as a result, preventing pollination.

Other weather and soil factors can hinder fruit set. Production is retarded when plants are excessively shaded. Also, under low soil moisture conditions, when water-stress in the tomato plant is most likely to occur, flowers often fail to produce a mature fruit. This condition may result from some interruption in the fertilization of the flower or to a disruption of growth of the little fruit following fertilization.

Q. I need counseling about nematodes, since my efforts to get rid of them have not paid off.

In preparing my vegetable bed, I watered the ground very well, put in VC-13, watered again, and waited two weeks before I planted my tomatoes and parsley. The tomatoes barely lasted three months, then turned yellow and died. When I pulled out the plants, the roots were full of "marbles" of all sizes. Same with my parsley, my flowers, and my papayas.

A. VC-13 is a good control of chinch bugs, but it doesn't worry nematodes. Try Nemagon (don't use on peppers). It is applied in furrows 12 inches apart and six inches deep. Cover the furrows, keep them wet for 48 hours, then let the ground aereate for four weeks. Sarolex, a form of Diazinon, is also a good nematode control, especially in lawns.

To restrain them in already established plots, fertilize heavily, applying three to four measuring tablespoons of fertilizer every 10 days to two weeks.

8. Annuals, Perennials, and Bulbs

For special index to this chapter, see back of book.

The tropics looks primarily to flowering trees, bold foliage plants, and brilliant shrubs and orchids for its color, so when flowers familiar to cooler areas are introduced, a little extra attention may be needed.

Because of soil-borne fungus and diseases, seeds are seldom sprouted in the garden. They usually are raised in sterilized soil in flats or containers, then set out when they have reached transplanting size. Or flowers are purchased as bedding plants from nurseries.

Planting seasons are different in the tropics (see the chapter on vegetables) and so are the varieties of flowers for summer and winter planting. Summer flowers are the sun lovers like marigolds and zinnias. Flowers that prefer the fall and winter here seldom withstand the warm, rainy weather.

The list of Northern-type perennials for the coastal tropics and semi-tropics is short, since many of the perennials grown farther north can't keep up with the long, warm growing season. Bulbs and tubers that do well farther north should be avoided as disappointments. Bulbs like Easter lilies and amaryllis, however, are beautiful substitutes.

To save money and effort, gardeners should consult their county agents and nurserymen before making planting plans or buying from seed catalogs.

Q. I would enjoy growing certain flowers from seed so I could afford splashier displays, but I just have no knack for sprouting much beyond marigold seeds here. (And, frankly, they usually sprout by themselves.) Any suggestions?

A. Garden shops sell gray plastic flats that are ideal for sprouting quantities of seed. For the medium, use peat moss and perlite, half-and-half.

Put this mix in your flat and water it well, checking to make sure the water has penetrated all the way. Then scatter your seeds across the top and just water them in.

Next, cover the surface of the flat with a single layer of gravel, preferably orange Chattahoochee rock although you can use white pea rock.

Set the flat off the ground on a bench or cement block under a tree or shrub that will give it dappled sun almost all day. Water it every other day or when the surface just begins to dry in spots. Do not overwater, and don't neglect it.

If you are terrified of damping-off (seedling collapse), you might want to water down the flat initially with Captan fungicide, and repeat the treatment every other watering.

Watering is crucial with many seeds since if a seed sprouts and then is allowed to dry out even briefly, it's usually doomed.

Relatively large seeds may be dropped into shallow holes poked in with a pencil and just barely covered. The hole should be no deeper than twice the depth of the seed.

The "stepchildren" in Florida gardens are the annual flowers that are given more prominence in colder climates. Because they require more effort than tropical flowering trees and shrubs, they are not widely planted here. Still, from October through May they will flourish.

Q We have a small flower garden (about 10 by 15 feet) in which we have planted marigold, zinnia, and aster seeds and a flat of petunias. Last year we had excellent results, but this year all the plants seem stunted. We used chlordane for red ants and a bag of composted cow manure. Is it the weather or what?

A. If there is no apparent sign of insect or disease damage, review your fertilizers. Adding manure to the planting bed is fine because it adds organic matter to the soil, but it is not a complete fertilizer.

You should have put down four to six pounds of a 6-6-6 garden fertilizer at the time you dug your planting bed. The usual rate is three to four pounds per 100 square feet of area.

A continuing program also is needed to push annual plants for best growth. A light application of fertilizer between the rows of plants every three weeks during the growing season should be adequate.

Q. I have a wonderful garden in the winter. My problem is in the summer when I don't plant anything. My garden plot is loaded with all kinds of weeds. Will you let me know what I could use to kill them and do no damage to my soil?

A. You could get a weed killer from a garden supply store and use the material as needed to keep down the weeds, checking the label for the period which must go by before it is safe to use the garden plot for plants.

A better way might be to mulch the plot rather heavily during the summer months, using grass clippings, leaves and other organic matter. This will help to keep down the weeds, and it also will add valuable organic material to the soil. In the fall, you can rake off the material that has not decayed and spade the bed as usual.

Q. I fell in love with gerberas when I saw how much larger and taller they grow here. I want to have a long bed of them along my driveway but so far I have had trouble keeping them healthy. Do they need special care?

A. Gerberas (African daisies) are targets for nematodes and soil-borne diseases, and the best preventative is to sterilize the bed with Vapam or Nemagon before planting. You can eliminate most damping-off problems without fumigating by using topsoil already sterilized. Gerberas need regular and frequent fertilizing too.

Q. My chrysanthemums have been growing like mad, but while the upper leaves look good, the lower leaves keep falling off. The plants look top-heavy. Any suggestions?

A. There's a leaf spot fungus that causes damage to chrysanthemums matching your description. A weekly spraying of Zineb or Maneb will control it. Try to feed your mums at least once a month to insure vigorous growth and plenty of blossoms in the fall.

Q. My zinnias' leaves are coated with a white film that I think is mildew. Is there a certain time of the day best to spray for this?

A. Don't wait until after 3:00 p.m. to water your plants and you'll help avoid powdery mildew. But spray for them any time with Karathane or Acti-Dione PM. Either will rout the fungus within a week or two.

Q. I brought home some beautiful-looking petunias in little plastic bedding pots. When I set them out the flowers were upright and the leaves crisp. From then on it was downhill — the plants wilted despite ample watering and I had to write them off. What was most likely the matter?

A. Assuming that the plants were growing in disease-free soil when you bought them, it could be that they were pot-bound. If they were large enough to be in full flower (more than just a few blooms) and the pots were undersize in proportion to the plant, it may have been too late to undo the damage caused by a crowded root system. Severely pot-bound annuals aren't as receptive to transplanting as are other plants and shrubs.

Petunias should be planted in an improved, well-drained soil. Fertilize with a complete mix such as 6-6-6. They like full sun or partial shade. During the hot summer months, they have a tendency to die. Your problem sounds very much as if the soil has nematodes. Use a soil fumigant about 10 days before planting.

They are also subject to crown rot which can be controlled with a solution of Captan or Fermate. Pinch out the center shoot to induce branching and avoid overwatering.

Q. What's the best control for spider mites? I have the little red devils all over my marigolds again this summer.

A. Generally speaking, miticides like Kelthane, Aramite, Tedion, and Chlorobenzilate are among the least toxic pesticides. A combination spray containing Ethion plus oil emulsion or Malathion plus oil emulsion is recommended by the County Cooperative Extension Service for excellent control.

Q. I thought when I got rid of the red spider mites on my marigolds I'd have healthy-looking flowers again but the blossoms continue to come out deformed. Or they simply fall off. What now?

A. When you spray for red spider, apply some Sevin, too. If you take apart one of those distorted blossoms you should find a corn earworm at work inside. Spraying with Kelthane or Ethion for the mites won't discourage him, but Sevin will.

Q. There are no seeds in my sunflower heads when they ripen, just hollow, fat husks. The last large head had dropped its pollen flowerettes when I tied nylon net over it to see if this would help produce seeds. Result: two cups of fat husks, nary a seed. Sunflower seeds are a very tasty nut for munching and, since these plants grow and produce such lavish heads with no production of seeds, I would like to solve the problem.

A. This is a common condition when sunflowers are planted on muck soils. Humidity is also a factor. They do better on sandy soil with less nitrogen content.

Q. I received several potted begonias for Christmas. Now the main stalks on some have turned dark and mushy and have fallen over and are dying. I have not changed the original soil in the pots or put on any fertilizer. I've only watered them moderately. What should I do to save them?

A. You can blame the mushy condition on a fungus, but this problem is encouraged by too much moisture. Hopefully, the drainage from the pots is good.

Water only when absolutely necessary. You might have to water them twice a week, but do it less often if you can get by with it.

Good air circulation around the plants will help. They won't do well in a really windy spot, but they do need circulation to keep them from being too moist.

You might try drenching the soil with Captan, a fungicide. Follow directions on the label.

Q. We have a seemingly very healthy bed of geraniums, but they do not bloom. Can you tell me how to make them flower?

A. First, consider the light received by the bed. Geraniums need full sun to bloom their best. In shade they may grow well but not flower.

Second, consider how you have fertilized the plants. Geraniums do not need a particularly rich soil. If overfed, they may produce plenty of leaves but fail to bloom. This situation would be aggravated if you have used high-nitrogen fertilizer like manures or Milorganite.

If you have been feeding the plants more than monthly, pay less attention to them. If you have used the high-nitrogen fertilizer, try another which is high in potash and low in nitrogen.

Finally, don't keep the plants wet. They prefer a fairly dry situation.

Overwatering may not be too much of a factor as far as bloom is concerned, but a moist situation may cause root rot in your plants. Under eaves with a southern exposure is ideal for geraniums.

Q. Why do I have so much trouble raising gloxinias? I purchase mine from a nursery but they never seem to thrive. What should I do?

A. The plants you buy are raised under very exacting greenhouse conditions which are really quite hard to duplicate.

They will grow in temperatures above 65 degrees and below 80. The air around them needs to be very humid. Shelter them from strong sunlight.

When flower buds are forming, apply liquid fertilizer twice weekly, but discontinue the feeding when flowers begin to open. You can prolong the life of the flower if you keep the gloxinias in a cool area. As the blooms begin to fade, cut back on the watering and when the leaves die, store the tubers in their pots tipped over on their sides.

Q. I had a hydrangea given to me with a blossom on it for Easter. Please tell me if this potted plant will do well if it is planted in a South Florida garden.

A. No, you could not expect the hydrangea to do well in the Miami area. It needs colder weather to bloom well. The plant is recommended for North Florida and parts of the central section.

Q. I transplant healthy periwinkles and for a time they appear healthy and vigorous but soon they dry up and die. I have not always had this trouble. Why now?

A. This often happens to periwinkles. It may be a soil organism such as fusarium. Best advice is to fumigate the soil or find another place to grow them.

Q. I don't see why other people can get my favorite plant to grow all over everything and it dies for me. I'm enclosing a sample from a periwinkle that turned brown and wilted, even though I gave it plenty of water.

A. It may be too much water that caused the canker and dieback that's your problem. This disease is always most prevalent during very rainy seasons. The shoot tips become dark brown, wilt, and die back to the soil surface. Some of the affected stems are nearly black in color.

Standard procedure to get rid of it is to remove infected plants or prune out diseased parts and burn them. Then spray with a copper fungicide three times at two-week intervals, starting when the new growth appears in spring. Be sure to coat the stems as well as the leaves.

Q. What's the season for caladiums in South Florida and should they be taken up when it's over?

A. Caladiums are a great source for color from March through late summer. Many people leave them in the ground. You can harden them off with a final feeding in September of a fertilizer high in potash (the last number on the fertilizer bag).

Caladiums here, generally speaking, like dappled sun or high shade (under pines or poincianas, for example). They should have a fast-draining soil with plenty of peat moss dug in. Water frequently and fertilize monthly with a 6-6-6 or composted manure.

Q. I've been given as Easter lily and would like a few tips on taking care of this and other bulb plants. Do you need to dig and store them here in South Florida?

A. Easter lilies like fertile soil in full sun and may bloom any time from February to April, depending on the weather. In Dade County, plantings are left undisturbed for several years.

Other bulbs that don't need to be dug and stored include daylilies, zephyranthes (rainlilies), the amaryllis, caladium, calla, crinum and canna lilies.

Bulbs ordinarily lifted after they flower and mature include gladioli and callas. These two may be left in the soil after they finish blooming if your ground is especially well drained, but you run the risk of decay during the wet season. Soggy ground will cause problems, though, even with caladium tubers so if you're in doubt, lift.

Before lifting gladiolus, be sure that the bulbs, or corms, are mature. New corms are made after the flower appears. Dig them when the top dies back, remove the outer husk, wash, and dry them. Put the corms in a dry, cool place (not your refrigerator) where there is good air circulation. A little peat moss may be placed in the bag with them.

Before they're replanted, gladiolus corms should be cured for a month to six weeks. If you are going to keep them out of the ground for several months after they are cured, you may have to store them in the lower part of your refrigerator. Otherwise, they are likely to begin sprouting if the atmosphere becomes moist.

Q. I need Easter lily advice. Mine get very tall (about 36 or 40 inches) but I get no buds or blooms on them.

A. Without knowing more about your plants and how they have been cared for, only speculation on a couple of points is possible.

Are your plants in the shade? Many flowering plants will grow well when shaded, but they may refuse to bloom unless they get sunlight — in the case of Easter lilies, full sun.

Easter lilies grow with ease in South Florida, where they usually are left in the ground year-round. Virtually pest-free, they are capable of producing more than 80 flowers on one plant.

Feeding the plants with a high-nitrogen fertilizer might contribute to the problem. A fertilizer relatively low in nitrogen and high in potash might help.

Q. My daylilies are infested with tiny black grasshoppers. Help!
A. This is not unusual, but you must take immediate steps or these immature lubbers will really deface the foliage. Dust with chlordane or spray with Sevin. Or knock them with the lid into a coffee can.

Q. Would you please offer suggestions on how to help Eucharis lilies bloom?
A. This plant seems to bloom best when grown in a rather crowded situation. Growing it in a pot or other container which restricts it will

Annuals, Perennials, and Bulbs **147**

achieve this. If you see a plant blooming well in the open, chances are it is an old, established clump.

The soil should have a good amount of organic matter, and the plant should be grown in fairly deep shade.

Q. I had a neighbor once who had a way of slicing up bulbs to make more bulbs. I've forgotton how she did it but I'd like to find out so I can make my beds longer.

A. With amaryllis, take a dormant or near-dormant bulb, trimming off all the roots and, with a sharp knife, make an X-shaped cut in the bottom, extending it a little over half way up the length of the bulb. Between each of the equal quarters formed, put in a pebble to keep them spread apart. Then bury the bulb in a pot with the top exposed but the cuts covered.

A good propagation medium is half sand, half peat. It should be kept moist, not wet. The small bulbs will come out between the "scales," and can be removed and repotted.

Bulbs like the Easter lily may have their scales carefully removed and inserted about a half-inch in a tray of rooting medium. In a few weeks the small bulbs will form, and when they are large enough to handle (about the end of the season) they can be planted in community pots.

These baby bulbs will take two or three seasons before they are ready to bloom, but they are an economical way to build spectacular beds.

Q. My amaryllis spikes had red streaks along the sides and were shorter than usual, curved, and stunted. There are rusty spots on the leaves. What is the problem and what could I have done about it?

A. "Fire blotch" is the name of the symptoms you describe. Someone who has never experienced it does not have to get excited over its possible appearance, but a gardener who finds it in the yard should take some precautions for the next season. (It's too late now.)

To prevent the fungus from taking over, buy or save only fat, healthy bulbs. Dip them for 10 minutes in a Benomyl solution (proportions on the bottle), then when the spike appears, spray weekly with Benomyl. If you can locate a material called Bravo, add this too, according to directions on the label.

Q. I have some big amaryllis blooming. Should I break the bloom off when the seed forms or let it stay on the plant?

A. If you are interested in seed propagation, by all means let the seed pod form. They develop rapidly and are mature within four to five weeks after the flower has been pollinated. Pods should be picked as soon as they turn yellow and begin to break open. Plant the seeds, after a few days of drying, in flats or beds in well-drained soil. Otherwise, you can "whack off" the bloom when it starts to fade.

Q. What do you do with calla lilies when the edges of the yellow flower part turn brown? Also, on a few of my plants, the leaves droop, die, and rot at the base of the lily.

A. You may have to squint to see them but the browning is probably due to thrips, very small, slender insects that like to suck plant juices. Chase them with Malathion, Dimethoate (Cygon, DeFend), or Meta-Systox-R.

The other problem is due to a bacteria or fungi that can only be prevented by cleaning the rhizomes before you plant them, cutting out any decayed spots. Let them dry a day or two, then soak them for an hour in two teaspoons of Ceresan plus a teaspoon of detergent in a gallon of water.

Q. Should a gloriosa lily bulb be taken up each year after the blooming season?

A. The usual rule on gloriosas is to dig and replant every few years. Digging annually isn't necessary.

Q. Among the plants we inherited from the former owner of our house was a bed of what she called "rain lilies." What do we need to know to keep these delicate little flowers thriving?

A. They look fragile, but these dainty amaryllids do well on their own in soil of moderate fertility and in either full sun or light shade. The miniatures — known also as zephyr, fairy, and prairie lilies — produce many paper thin seeds but multiply rapidly by bulb, too. A once-a-year spring fertilizing and adequate weeding will keep them producing season after season. They also are called zephyranthes.

Q. My dahlias have produced beautiful flowers. Now I would like to know what to do since they have finished blooming?

A. The clumps should be carefully dug up, cleaned, and stored in dry peat. Dahlias have large tubers and each must have with it a piece of the stem upon which a healthy bud can be seen. If they are broken off at the neck as a result of careless handling they will be worthless.

These flowers ordinarily do not do well in South Florida.

Q. I like to keep fresh flowers in the house but during the warm weather they often wilt quickly. We rarely use our air conditioner, but always have good ventilation with fans. I have tried aspirin and other unusual ideas but with no success. Would appreciate any suggestions.

A. Have you tried "conditioning?" Many flowers benefit from overnight conditioning. Chrysanthemums, lantana, oleander, salvia, and calendulas, and flowers with hollow stems (like hollyhock), herbaceous woody stems (lantana), or hairy stems (calendulas) keep better if first placed in moderately hot water.

The water will cool, of course, but let them remain in it overnight in a cool place. Split stems of gaillardias and chrysanthemums before conditioning. Add three tablespoons of sugar to each quart of water for chrysanthemums, and remove as few leaves as possible above the water line.

If not completely wilted, many flowers can be revived by reconditioning.

Q. What do you do with cannas that look like mine? The leaves are ragged, jagged and chewed up, and folded over. I can't see what's causing the damage.

A. Look again — this time inside those folds. You'll probably find a long, grayish-green worm with a large brown head. This is the canna leaf roller, and he isn't satisfied unless he can put a leaf under his belt every two or three days.

Controlling him isn't easy; it's best to keep a stomach poison on the leaves throughout the summer months. Try Sevin.

Q. We would like to try planting some iris. Could you give us any suggestions on variety, location, and care?

A. There are seven species of iris native in Florida. Iris from Louisiana, as well, are good garden perennials, but forget about the German types that you may have known and loved in the North. Iris like soil rich in humus and plenty of water during dry periods. You can fertilize during the summer with a 6-6-6 fertilizer mix at the rate of five pounds per one hundred square feet of bed. Soil sterilization is strongly recommended.

Plant the bulbs about four inches deep in October. You can also plant them in pots, six bulbs in a six-inch pot. Until green tips appear the pot should be kept in the dark and only moved to full light after the tips are about a half-inch above the soil.

Q. I'd like to be able to dry some of the flowers I raise. What is the best method, and can I dry leaves successfully, too?

A. The old-fashioned technique of hanging for small, compact blooms is not always effective in Florida, depending on the dampness in the air. For those afraid of attracting bugs, the meal and borax treatment has less appeal than silica gel. Less expensive is sand and borax. Glycerin-and-water-treated leaves will retain their flexibility but may undergo a color change. Pressing leaves will retain the color, however they will be flat and brittle. The former method is best for an arrangement. Large leaves (like palms) and flowering stalks may be left in a pail of 100 percent glycerin for a month or more.

9. Orchids, Bromeliads, and Ferns

For special index to this chapter, see back of book.

Some can be grown in the ground, some in the air, but bromelaids, ferns, and orchids are being grown everywhere today, enjoying peak popularity around both homes and apartments.

Comparison shopping will show that gorgeous blooming orchids are available at a wide range of prices — from a couple of dollars to $10,000. The best place to find bargains is through a small dealer or hobbyist advertising in the classified under "Plants." Then you won't feel your first ones need to be handled with kid gloves. Orchids are sturdier and more resilient to problems than many people believe — if they are first provided with good air circulation and water that drains rapidly. Public libraries have plenty of material on growing them to perfection.

Bromeliads are even tougher, and many have foliage and flowers just as brilliant. They're unusual and unusually resistant to problems where air circulation is good and humidity is high. They also propagate themselves eagerly for attractive displays as borders and beds.

Q. I am very eager to collect more orchids and bromeliads, but I live in a relatively new development where the trees are not large enough yet to provide the right natural setting for ferns, air plants, and the kind of jungle look I want. Short of building a shade house — which, I'm afraid, would make our yard look even more stark — is there any temporary device or structure I can use on a small scale so I can start collecting?

A. Here is one approach you might use that provides natural shade quickly.

To give you, over a relatively large area in a relatively short time, the kind of dappled sun most ferns, orchids, and bromeliads love, you could

Many ferns are native to South Florida but scores of introduced plants thrive as well. Maidenhair is a favorite choice of gardeners with waterfalls; it enjoys shade and high humidity.

Orchids, Bromeliads, and Ferns **153**

quickly and cheaply construct a sort of arbor. About 12 feet or so from an exterior wall sink a rough cedar four-by-four a couple of feet in the ground in cement. Fasten a large screw eye in the top, then measure off about 12 feet along the eaves of your roof. Screw into the wood or cement into the wall more screw eyes eight inches to a foot apart.

Then fasten nylon cords from the big screw eye on the pole to the screw eyes along the roof.

Plant a fast-growing vine at the base of the pole in a good-sized hole filled with equal parts peat moss and topsoil, mixed with a coffee can of sheep manure or other organic.

Good choices for the vine would be *Thunbergia erecta,* which has big blue trumpet flowers with yellow throats, and one of the passion vines. Both are very fast growers and will cover the arbor in just a few months if watered regularly. It's easy to prune them to get just as much sun as you want.

Q. How do you fasten orchids onto a tree, what kind of sun should they have and how often should you water them?

A. Take your plant out of the pot, leave any osmunda on the roots and bind it securely upright with new growths close to the tree so that the roots will grow more quickly onto the trunk. The plant must be firmly wired so that the tender root tips will not be broken by an accidental nudge or by slippage.

Locate orchids so they have good light but filtered through tree leaves. (Vandaceous types, however, can take more sun.) Until the plant develops a well-established root system over the bark of the tree, you'll have to water it frequently, letting it become bone-dry between waterings. Afterward, except during droughts, nature should provide enough moisture.

Q. I learned the hard way that water softener is severely injurious to orchids. Could you tell me if any other plants are sensitive to it?

A. Sounds like you've fallen prey to a rumor-monger. The experts have never heard of damage to orchids or any other plants by an ordinary water softener (unless the timer has been affected by power shortages so it discharges chlorides). Cases are known where people have disasterously added agents to tap water to increase its acidity (hopefully, to make some nutrients easier for the plant to assimilate).

You may have attributed the decline of your plants to the wrong source. Overwatering or underwatering, or too much or too little fertilizer, are just a few of the excesses or deficiencies that can affect orchids.

Q. How often do you fertilize cattleyas and what do you use?

A. About once every six weeks during the summer using Peters 20-20-20 in a sprayer. An ideal orchid house for cattleyas has a fiberglass roof with Saran sides that gives plenty of air movement and humidity but protection from heavy rains.

Q. What do you grow paphiopedilums in?
A. A mixture of "Black Magic" with fir bark and tree fern (moisture retentive yet conducive to good drainage) in plastic pots is preferred. Also recommended is a mix of bark, tree fern and oak leaf mold. Repotting is unnecessary until the plant has filled the pot. Generally, avoid adding lime of any kind to the mix.

Q. In what medium do you pot cattleyas, and why, when and how do you repot?
A. A mixture of three parts tree fern and one part redwood chips is standard. Redwood acidifies the medium, minimizing possibility of fungus activity. Tree fern drains well and has good nutrient absorption characteristics. Repotting is usually necessary after two years as decomposition of the medium begins to take place.

Q. I have a phalaenopsis orchid which is blooming for the first time, and all the buds continue to turn a pale yellow and fall off. It's growing in bark. Just what is my problem?
A. If the leaves aren't spotted, your plant is probably getting too much or too little water or fertilizer. A leaf spot fungus could also cause the bud problem. Many growers are using Benlate, a very effective material. Or you can use the standby remedy of soaking your plants for 60 minutes in natriphene or 8-hydroxy quinolin.

Q. How often do you fertilize phalaenopsis?
A. Use high nitrogen (30-10-10) in the summer when the plants are growing, fertilizing every ten days to two weeks. In the winter, use 20-20-20 every three to four weeks and saturate the pots.

Q. How often do you water phalaenopsis, how do you know it's time to water and how much do you give them?
A. Water two or three times a week, depending on the weather. If you touch the outside of the phalaenopsis pots with your hand and they feel room temperature, this indicated dryness. Morning watering is always best. Water thoroughly but withhold somewhat in wet weather.

Phalaenopsis in glass houses usually are given 60 percent shade. Humidity is controlled by using fans year-round.

Phalaenopsis — the moth orchids — grow spectacularly hung from trees which provide all-day filtered sun. They thrive on plenty of water and little fertilizing, although regular feeding produces a greater abundance of flowers.

Q. How should you grow dendrobiums?

A. Many hobby growers have theirs on patios under 40 percent shade. They do not control temperature or humidity, and the prevailing breezes provide adequate air movement.

Q. What do you pot dendrobiums in and how often do you repot? What fertilizer should be used, and when?

A. Pot in 80 percent tree fern with 20 percent redwood bark chips — tree fern for the heavy rains, chips to help control pH and hold down mold. Repot every two or three years, only when new growth starts.

Most growers fertilize their plants weekly when they're in active growth and every two or three weeks after blooming. Use Peters 20-20-20, and wet the foliage and pot well.

Q. Under what conditions do vandas grow best?

A. Different types grow under a variety of conditions, including "wild" with up to 60 percent shade.

Some hobbyists use 52 percent Saran cloth for plenty of sun and just enough shade for vandas to bloom well. During hot spells, water more often, but otherwise temperature and humidity do not have to be controlled on a regular basis. Sixty five percent Saran cloth and fans in very hot weather are preferred by some. Vanda 'Miss Joaquim' grows in full sun.

Various potting media are used by hobby growers, including Solite, firbark, Gore mixture, and combinations. Water every other day, usually in the morning.

Q. What fungicides and insecticides do you use on oncidiums?

A. Captan or Phaltan in solution are used usually after several rainy days. Apply Sevin about once a month. Red spiders do not seem to affect oncidiums, but they get a treatment of Dimite along with the other orchids once or twice a year from many growers.

Scale is treated on individual plants with a Malathion solution, scrubbed on with a soft toothbrush. You might want to use chlordane for ants when needed and roach pellets.

Q. I have an orchid that won't behave. It keeps throwing its roots outside of its pot instead of down into the potting material. Why this strange behavior?

A. Orchids occasionally throw roots into the air in memory of their epiphyte heritage, but too many outside the pot indicates something is wrong inside. Your plants probably need repotting, either because the potting material is too old and breaking down, or because you have been overwatering and have turned the potting material sour and caused the roots inside to rot.

Q. I found a bromeliad in a tree, took it home, and it did well, but now I notice it's getting that hungry look. How do you feed these plants?

Like most bromeliads, *Neoreglia carolinae* var. tricolor — a yellow-and-green-striped beauty with a rose center — can be wired to driftwood and will thrive if its center "cup" is kept full of water. Even without fertilizing these plants will grow under trees, where falling leaves catch in the cup, decaying and feeding the plants.

A. A liquid fertilizer like those used for orchids will do the job. The amount depends on the strength of light your plant receives. In strong light, once-a-week fertilizing to the cup and roots may be necessary. Once every two or three weeks is appropriate for bromeliads in reduced light. Most important is frequent watering.

Q. I would like some information on how to plant bromeliads. Should they be potted or planted in open ground?
A. These plants have great variation in size, shape, and foliage color. They grow well indoors and out. They seem to do best in very porous organic medium such as equal parts of peat moss, leaf mold, and sand. Chopped osmunda, bark, or tree fern can be used in place of the peat moss.

Q. I have two bromeliads that I acquired recently under the impression that these plants seldom have any complications. Now one plant has small, hard black dots on it, and the other is developing a gray scale. What's the matter?
A. Bad luck in the form of flyspeck scale and palm scale. Bromeliad raisers very seldom have to take action against these pests but when they appear in considerable numbers, a tablespoon of Malathion (50 percent

emulsion) to a gallon of water is the best control. Dip the entire plant and let a film of the solution remain on it overnight. The dead scale will have to be removed with a knife.

Q. Why can't I transplant ferns successfully? I'm not trying anything delicate — just some ferns that grow wild all over my neighbor's yard. She doesn't give them any special soil or water. For the first few weeks after I transplant them I water them every day but they still turn brown and die.
A. It hurts, but ferns should be given a "haircut" when they are pulled up for transplanting. Cut all of the stems (except for tiny emerging fronds) back to an inch in height. This gives added vigor to a plant and makes readjustment much easier.

It is possible to leave a couple of short fronds on a fern wrenched out of its original home and see them survive if plenty of humus or peat moss — not to mention water — will be available at the new bed.

Almost needless to mention is the fact that few ferns can survive a too-sunny spot in hot weather when their root systems have been disrupted.

Q. What do you use as spray for mealybugs? I have them on my fishtail fern.
A. Use powdered Malathion, Sevin, or soapy water. Look around for ants. They usually carry mealybugs to the plants. Get rid of the ants and very likely you will get rid of the mealybugs. This is mainly an indoor plant problem.

Q. How often should I fertilize?
A. That's up to you. How fast do you want growth? Fertilize regularly — at least once a week — and you will get a beautiful plant. When you are feeding this often, use one teaspoon to a gallon of water. Don't overfertilize because fertilizers contain salt and you can have salt burn on your ferns.

Q. What about feeding ferns in hanging baskets?
A. Make up a sufficient solution of 20-20-20 in a wide bucket and immerse the basket in this solution about to the rim.

Q. What about slugs on ferns?
A. Use a bait containing metaldehyde. Do not place it on the crown of ferns or in contact with new growing rhizome tips. If your pots are off the ground, the problem won't be nearly as bad.

Q. Is there any spray for worms on ferns? I squashed one but there is still something eating it.

A. You can use Sevin as a wettable powder or dust. The five percent dust or the 40-50 percent wettable powder are both satisfactory for home use. If you use the dust, do not wet your plant before you dust it.

Of course, never spray or dust or even fertilize any plant that is in a wilted state. Water and wait until the plant perks up. By no means use liquid Sevin. It has an oil base and should be avoided. Whatever insecticide you use, do read the instructions, even if you think you remembered reading them before. (The little green worms are known as lupers and the brown worms as cutworms.)

Q. If you have a variety of ferns, can you water them all the same way?

A. No. How often you water each depends on many factors such as the type of fern (epiphytic or terrestrial), the size of the plant, the rhizome, kind of medium, the type of container, and the size, plus the location.

For most ferns it is suggested that you feel the medium (the soil). If it feels wet, it does not need watering. If it feels dry, it should be watered, and if it is in a wilted state, it should have been watered yesterday.

Q. What is a good medium for bird's-nest ferns?

A. Since this is an epiphytic fern growing on trees or rocks, you do not need to use a strictly soil medium. In South Florida it grows on limestone rocks. Use light, organic-type materials — orchid fern bark, osmunda, pine bark, and inch-square chunks of tree fern. Do not use as much perlite and lime rock screenings as you would for your terrestrial-growing plants.

There are bird's-nests that have done beautifully for years and then suddenly started going to pot. Probably the trouble has been that the potting medium has broken down so that it no longer drains properly. Since this is an epiphytic fern, it loves to have good drainage around its root system. When the medium breaks down, it gets soggy and does not drain properly, thus blocking the necessary aeration. The roots begin to die off. Solution: repotting.

Q. I'm confused. What's the wild Boston fern doing growing down here, and does it like sun or shade?

A. The fern is native to South Florida, the West Indies, Bermuda, Queensland (Australia), and other tropical climes. It was introduced to the North about 70 years ago as a houseplant, and evidently was particularly welcome in Boston parlors. It likes shade, but if given plenty of water, this excellent, easy-care ground cover can be grown in full sun.

Q. What is tree fern fiber?
A. It's the fibrous, woody trunk of a tree fern and it is available at any shop or nursery selling orchid supplies. It is available in many forms — in slabs, carved monkeys or baskets, and in three or four shredded grades used as orchid medium and sometimes in fern and other mixes. It holds water yet provides plenty of air circulation.

10. Indoor and Terrace Plants

For special index to this chapter, see back of book.

Until just recently when condominiums and high-rise apartments became a significant part of the landscape, there was very little interest in South Florida in house or terrace plants. But the enormous increase in gardening since the early 1970s has infiltrated nearly every multiple-unit residence and has created a considerable demand for information on growing tropical plants of all kinds.

Nearly everyone realizes that these plants must have special conditions created for them within these artificial environments which are drier and dimmer than most of them prefer. There are many recent books available on plant selection and care in apartments, and this book does not attempt to cover the subject. It does give some of the most frequently asked questions of general interest here.

For further help, the following books are recommended: *The Apartment Gardener* by Florence and Stanley Dworkin (Signet paperback), *The Complete Book of Houseplants* by Charles Marden Fitch (Hawthorn), *House Plants Indoors/Outdoors* (Ortho), *Exotic House Plants* by A. B. Graf (Roehrs), *Hanging Plants for Modern Living* and *Foliage Plants for Modern Living,* both by M. Jane Coleman (Merchants Publishing Co.).

Q. How can you tell if what's wrong with your plant is from overwatering? I have a hard time keeping the soil moisture just right.

A. The symptoms of overwatering are many and frustrating. A plant that isn't growing at all but is in constantly moist soil is probably overwatered. Dropping of buds and flowers can come from overwatering. Rotting is one of the more obvious symptoms.

When several leaves turn yellow and fall, the most likely cause is overwatering. Brown margins or spots on leaves can be from overwatering. They also can be from overfertilizing.

No indoor plant should be potted in soil but should have a fast-draining, half peat moss, half perlite (those little white "squishable" rocks), or African violet mix. If your plant came potted in regular soil, repot.

Every pot should have a drainage hole. Although some recommend putting a layer of gravel in a pot that doesn't, the pros "double-pot" instead. They take the original plastic pot and set it on gravel inside a decorative pot. The gravel is extra insurance that the pot will not sit in water; however, you should learn to judge how much water is needed to soak the pot thoroughly without too much draining out. Or you should take the plastic pot out to water it.

To correctly water most plants except cacti and succulents, soak the soil thoroughly so water comes out the holes. Then do not water again until the top quarter-inch of the soil feels dry — not just cool or not just looks dry.

Overwatering is the most common cause of potted plant sickness.

Balcony gardens can look lush year-round in Florida if plants that are suitable for up-in-the-air culture are selected. Common landscaping plants often can be used to serve as a buffer against drying wind for more tender exotic specimens.

Q. I live in an apartment with few windows and the space around those is limited. I'd like to have some plants indoors and can offer them plenty of artificial light, but is this enough?

A. You can have an indoor garden located in a closet if there's a source of ventilation. Almost any plant can be grown under artificial light, although some fare better than others. African violets, gloxinias, philodendrons, some begonias, and orchids take especially well to this type of cultivation.

Ordinary light bulbs aren't adequate, however. They are too hot, weak in red or blue "color bands" found in natural light, and costly, delivering only a third as much light for the money as do flourescent lights. While the so-called grow-lights have been recommended in recent years, U.S.D.A. researchers now say flourescent lights produce as good or better plants.

Q. Is it impossible to grow a tree on an apartment terrace? (It has a southern exposure.) I'm trying it but the plant doesn't look well. Many of the leaves are dying back from the tips, even though it gets sun most of the day and is well watered (but not soggy) and fertilized regularly. I recently repotted it. Could there be something wrong with the container?

A. There are six main areas to investigate when a plant has "tip burn," the condition you describe.

1. Pot-bound roots. If the tree was planted in a too-small container or if it had outgrown the original pot, this could cause leaf damage.

2. Root damage. Improper planting may have damaged the roots.

3. An aeration problem. Not many trees like compacted soil. Repotting in a half-sand, half-peat mixture, well-fertilized, would help correct this.

4. A container treated with a harmful chemical. If the pot was not constructed for use as a planter, it may have some chemical residue in it that could affect the tree.

5. Overfertilizing. Plants in containers need less frequent fertilizing than outdoor trees.

6. Location. A plant backed into a corner may be the victim of heat build-up as the sunlight reflects from the floor and surrounding walls. An air conditioner blowing on a plant in a pot will also cause rapid transpiration, which calls for much more frequent watering than you might expect.

If you want to double-check your soil, you might send a sample of your potting mixture to your County Cooperative Extension Service.

Q. I have a problem getting my African violet to bloom. I have had one plant for three years and it bloomed profusely last year. This year the buds came out and then died before opening. The plants are

healthy-looking and have good color. Can you give me some advice — or shall I throw them out and start over?

A. Without knowing more about the plant it's hard to give you a definite answer, but here are some suggestions based on the most common causes for failure to bloom.

We'll assume that you have continued the same feeding and watering routine you maintained last year when the plant blossomed and that you have not moved it to a darker spot or that its window is not receiving unaccustomed shade.

The plant should be carefully examined for insects. You might try a fertilizer higher in phosphate and potash than nitrogen, such as a 10-30-20.

If the plant appears somewhat wilted and dehydrated but the soil is adequately watered, the cause could be pot-bound roots. If your plant is in an air-conditioned room, changing temperature could be a factor. And, as a rule, remember that underwatering is preferable to overwatering African violets.

Q. We're going away for a couple of weeks and, being newcomers, don't know anyone well enough to ask them to take care of our house plants while we're gone. What do other people do when they're away? We've got some beauties and don't want to lose them.

A. Water them thoroughly before you leave and put them out of direct sunlight and drafts. Cover the soil with well-moistened cotton or commercially prepared peat moss, sphagnum moss, or leaf mold. Then group the plants together.

Here are some more ways indoor gardeners keep their plants moist while they're out of town:

Encase the plant and pot in a plastic polyethylene bag 12 hours after watering it thoroughly. Tie the bag loosely and punch a few holes in the top. Make certain that the plant does not touch the plastic bag by using stakes taller than the plant to support the bag.

Fill your bathtub with water and cover the surface with a mesh screen. The bathroom should have adequate lighting. Put about two inches of glass-fiber wick in the hole of the flower pot. Rest the pots on top of the mesh and lower the other end of the wick into the water. The wick soaks up water as needed by the plant.

Two or three weeks before leaving, either plant the pots outside in a shady location or completely transplant the plant in your garden. Water thoroughly before leaving.

Automatic watering gadgets can be purchased in several styles and prices that will quench your plants' thirst for a week or 10 days.

Q. I have a terrace on the fourteenth floor and nothing seems to grow well for me. It is very windy up here and things dry out fast. What few plants I've managed to hang onto look under par.

A. You might like to take a look at Edwin A. Menninger's *Seaside Plants of the World* at your library or bookshop. Whether you're on the water or not, you need plants that do not have easily dehydrated foliage.

Some of the desert plants would do well for you — some of the spineless dwarf agaves (miniature "century plants") and the night-blooming cereus, which could be trained along a railing or wall.

A durable hanging basket choice might be the yellow-flowered wedelia that is so often used as ground cover in South Florida. Also, the weeping purple lantanas are fairly rugged. The red, pink, yellow, or white flowering crown-of-thorns could be tucked into a corner in a pot. Periwinkles take to terraces and so does the white-flowering carissa called 'Boxwood Beauty.'

Sansevieria — especially the yellow-green variegated kind — look well in containers and will put up with almost any neglect. So will the bromeliad-like *Rhoeo spathacea,* the oyster plant. The variegated *Pittosporum tobira* is sturdy and attractive.

Generally speaking, you are going to find woody landscaping plants more suitable than the foliage plants sold for indoor use — or even outdoor use where there isn't an almost constant wind.

Q. I've just moved into a home that has a lovely screened room with an opening in the west corner of the cement floor. The room faces north and the plot of earth is 30 inches by 30 inches and rounded on the third side. What do you suggest I plant in it that doesn't require much sun and is easy to grow?

A. Dracaenas will tolerate a considerable amount of shade, as will dieffenbachias, a number of philodendrons, several ferns, Swiss-cheese plant, spathiphyllum, anthurium, aspidistra, sansevieria, and chamaedorea palms.

Plants of varied heights are included in the list. You may want to choose a tall, a medium, and a low plant for a grouping in space you have available.

Q. I don't have much luck with indoor plants. They look fine for a few months, then develop various complications, mostly dieback and some sort of withering ailment. One of my friends has a veritable jungle indoors but claims to do "nothing special" to her collection. Where could I be going wrong?

A. One way to grow indoor plants is to buy or propagate a corresponding set of "outdoor" plants. Most of the time, when an indoor plant begins to look sickly, loses its color, and starts to turn brown around the edges, all it takes is a month's vacation outside in the natural humidity it loves with partial or filtered sunlight to revive it. So set it outside and replace it with its stand-in from your "swing shift" collection.

If you're in an apartment where this is impossible — and especially if you use air conditioning a great deal — you may have to start a weekly, or even twice-weekly, schedule of leaf-washing or misting with clear water to help maintain a humid atmosphere.

Also, warm, dry, air-conditioning air encourages mealybugs, thrips, scales, and mites that can be curbed with spray from your garden supply shop. Remember, too, to keep up with your plant's growth rate by repotting.

Q. I live on a busy street and since I keep the windows open most of the time my houseplants — especially the dracaenas and monsteras — collect dust quickly. I use a soft, damp cloth to clean them but have been told that a plant polish would be better. I have always felt that giving plants an artifical gloss would be harmful. Am I wrong?

A. Most of these plant polishes leave an oily film that attracts dust but, at any rate, does not repel it. What you wind up with is a dusty polished surface instead of a dusty natural surface. Most plant lovers agree that the artificial look given by the polish — as intriguing as it might be at first — is not offset by any real advantages.

Q. What do I do when I want to put a plant in a pot into a hanging basket instead? Can I use ordinary potting soil?

A. "Ordinary potting soil" covers a wide range of sins. Potting soil can be simple 'glades marl, a poor choice for baskets or pots. Good drainage — which means water should not sit on the surface after you're through sprinkling it — is the most important factor. A satisfactory hanging basket mix with components that are easy to find would be equal parts peat moss, composted cow manure, and perlite.

If the basket is not plastic and is to be lined with sphagnum moss, don't stint with the moss, even though it is expensive. A skimpily lined basket dries out and deteriorates very fast. Soak the moss in a bucket of water and press it well into and through the bottoms and sides so that, compressed, it is at least an inch thick.

Q. During the past year, I've gone hanging basket crazy. I now have well over a hundred of them hanging in mango trees in my backyard. The thought occurs to me that obviously some of my plants will not

tolerate the cold and windy weather as well as others. Could you list some that need protection and what form this protection should take?

A. When the thermometer drops below the mid-fifties, it's time to take episcias and any of the gloxinia-gesneriad family indoors. Plastic coverings aren't enough. In the low forties and upper thirties damage can be considerable with maidenhair (but not Boston fern), impatiens, crossandra, peperomias, marantas, calantheas, some philodendrons, alocasias, dieffenbachias, and coleus, among others.

Orchid plants especially susceptible to cold weather, according to orchid growers Jones and Scully, Inc., are angraecum, brassia, catasetum, cycnoches, Indian and phalaenopsis-type dendrobiums, grammangis, grammatophyllum, all species and hybrids of phalaenopsis, schombocattleya "Angel Wings," and schomburgkia. In addition, some growers coddle *Epidendrum atropurpureum* in cold weather.

Ivies aren't hurt; neither are any other plants native to strictly temperate zone climates.

Anyone who collects hanging basket or potted specimens seriously should not take a chance when predicted temperatures are below 45. Drying winds often do more damage than the cold would alone, and no gardener can be certain that his neighborhood won't have temperatures below those predicted.

The ideal arrangement is to have a slat house that can be covered with plastic sheeting. Second-best is to stud garage or utility room walls with hooks so plants can be moved indoors. If neither move is possible, plastic from the hardware store can be stretched and nailed to cypress lath frames that can be positioned against a wall or fence to create a lean-to shelter under which exotics can be moved. Drench the ground underneath thoroughly first. Evaporation of the water during the night will help raise temperatures.

11. *Propagation*

*For special index to this chapter,
see back of book.*

In a tropical climate it doesn't take much time to propagate enough plants to make a lush showing around your property. There's no substitute for massed displays in a garden — whether you're concerned with spreading ground cover over a wide area or making a big splash with color in generous-sized beds.

It doesn't take as much time, either, as some might think, to grow good-sized plants from cuttings. The key is consistent moisture and protection against fungus. This can be accomplished with "clean" new potting soil and a soil drench of fungicide. The combination of peat moss and perlite, kept moist and in the shade, will root thousands of different plants, and these materials are almost universally obtainable at reasonable cost. Compare the cost of plants today, and you'll agree there's nothing you can "make" at home that will save as much money for you as do-it-yourself landscaping materials.

Q. What is the best way to make a miniature greenhouse to root leaf and stem cuttings of sensitive plants like begonias?

A. Clear plastic shoe boxes, large refrigerator boxes, and bread boxes are all good, filled with about 1½″ of sterile vermiculite or perlite. Soak the mix thoroughly and poke it to be sure. Also be sure excess water is tipped out.

Stick your cuttings into the medium so they stand upright, not quite touching the bottom, or pin rhizomes and leaf cuttings to the surface with small, bent pieces of wire or sterile pebbles. Close the top and set the container under a tree with not too heavy a canopy so that it gets dappled light. Some cuttings will root in as short a time as three days; others take much longer.

When they are large enough to transplant, keep a plastic bag over the new potted-up plants until they show obvious signs of growth.

Q. I have had no trouble rooting spider plants from the runners they send out, but I can't root runners from strawberries and saxifrage. I cut off the little plants, pot them up, keep them moist, and they die. What could I be doing wrong?

A. With the chlorophytum (spider plant), roots are usually well formed midair when the plants (stolons) are cut off, so they have a much easier time getting started. With episcias, saxifrage (strawberry begonia), and other plants that do not have well formed roots, it is better to move little pots up under the stolons and let them root while attached to the mother plant.

Sprouting ornamentals from seed is easy if seeds are planted no deeper than their width and the soil is kept moist but not soggy. One of the surest ways for most people to do this is to mix peat moss and perlite in equal measures in a plastic (not clay) pot, drench it with water, press seeds barely into the surface and cover with a single layer of gravel as mulch. Place under a tree and water daily.

The three stages of air layering, the most certain way to propagate many woody ornamentals: (1) an inch or two of bark is cut away in a ring from a branch, usually about 2 to 3 feet long; (2) the bare strip is wrapped in wet sphagnum moss and bound with a plastic wrap; (3) when roots show through extensively, the branch is cut off just below the wrap. The plastic is then removed and the branch is potted in a peat moss, perlite, and sand mix and kept moist.

Q. I would like some information on when is the best time in Florida to air layer plants — bottle-brush and orchid trees, especially. Does the kind of plant make any difference?

A. Spring and early summer are the suggested times because most plants will root faster at that time of year. Both the plants you mention are grown by seeds and by air layering.

A wide variety of woody plants may be grown by air layering, but there are some which are difficult. It's best to check a reference book or an expert if you are in doubt about a particular plant.

Q. Can air layered plants be removed from their parent plant and placed directly in the garden, or should they receive special treatment at first?

A. An ideal way to introduce them to soil would be to pot them in a peat-perlite-sand mixture kept moist until their limited root system takes hold. But they can go directly into soil in your yard if you pay close attention to watering and provide light shade to protect them from hot sun.

Newly rooted air layers require tender handling. After removal from stock tree and while potting them, take great care not to break off the roots that originate from the cells near or at the surface of the callous tissue. More air layered plants are lost from improper handling than anything else.

Q. How best can I propagate allamanda? I have tried layering to no avail. Does it root readily in soil, water, or how? And how does one cut a *Dracaena marginata* in order to induce a side shoot?

A. You should be able to grow allamanda rather easily from cuttings. You could use a rooting medium of perlite and peat moss.

A plastic cover or tent over the container in which you root the cuttings should reduce the loss of moisture. Keep the container in a light place, but not in direct sun. Heat from the sun on plastic would cook the cuttings.

If you need only a few plants and have a parent plant to work with, you can grow the plant by air layering.

The easiest answer for the dracaena is to cut off the top at a point where you want growth. You should get at least two shoots below that point. If you don't want to remove the top, you could try notching the bark just above where you want the new growth.

Some growers tie the main stem into an almost horizontal position and get new growth to start along the stem by that method.

Q. I would like to root nonblooming orange jasmine from cuttings. Should this be done in sand, rich soil, or water? About how long does it take for them to root?

A. For a rooting medium that is clean, easy to work with, and moisture retentive, you could use a mixture of perlite and peat moss. Covering the flat with plastic will help retain moisture and speed the rooting. The cuttings should root in a matter of a few weeks. Actually, you can grow this plant from seeds and by air layering as well as cuttings.

Q. I have trouble getting cuttings to root in my yard, mostly because I have very little shade. I start them in my small shady section, but when I transplant the young plants, they simply won't take although I water them liberally. Any sure-fire way to move them out with the rest of my plants?

A. Have you tried peat pots? Fun to work with, they are an easy way to get tender cuttings gradually used to full sun. Plant your cuttings in the pots, leave them a week in three-fourths sun, then give them the full treatment, planting the pots in the earth in full sun. This way there's no transplanting "trauma," and the peat pot will decompose and form plant food in a few short weeks.

Q. What can be done — if anything — to stop what I call root rot in seedlings?

A. The best control of this condition — called damping-off — is to treat the soil before planting your seeds. The fungi that causes damping-off can be squelched with a fungicide applied according to instructions. Annual and vegetable seeds dusted with a seed protectant such as Arasan or Captan will have a better chance for survival, too.

Sometimes drenching plants that are dying from damping-off is helpful. Terrachlor applied at the rate of one pound to each three gallons of water and poured over 1,000 feet of row, or neutral copper applied at the rate of a pound to three gallons of water over 1,000 feet is effective. Ferbam or

Captan may be used at a rate of two tablespoons to a gallon of water for 100 feet of row.

Q. When you're grafting, where should you cut off the rootstock and how thick should the stem of your introduction be?

A. The rootstock (the young plant grown especially for its known strength) is cut off about six inches from the ground. The scion (the choice piece you're going to graft onto the rootstock) should be the same width as the rootstock where it was cut. Of course, you can graft onto a branch of an older plant, keeping in mind that the only branch that will produce the new variety is the one with the graft. Joining pencil-sized widths is usually recommended.

In wedge grafting, you make about an inch-and-a-half long sloping cut into the rootstock. The lower end of the scion is cut into a wedge and inserted into the slot so that the entire exposed end of the wedge is enclosed and the cambium of stock and scion meet. Bind them with plastic. It'll take about six weeks, more or less, for the join to be made, depending on the type of plant, the weather, and other factors.

Bibliography

American Hibiscus Society. *What Every Hibiscus Grower Should Know.* Vero Beach, Fla., A.H.S. Publication Committee, 1974. Available through Fairchild Tropical Garden bookshop (10901 Old Cutler Road, Coral Gables, Florida 33156).

Brown, F. B. *Florida's Beautiful Crotons.* Indialantic, Fla.: B. F. Brown, 1961. Available from the author at 244 Michigan Avenue, Indialantic, or in bookshops.

Bush, C. S., and Morton, Julia F. *Native Trees and Plants for Florida Landscaping.* Bulletin 193. Tallahassee, Fla.: Florida State Department of Agriculture and Consumer Services, 1969.

Coleman, M. Jane. *Foliage Plants for Modern Living.* Kalamazoo, Mich.: Merchants Publishing Co., 1975.

Coleman, M. Jane. *Hanging Plants for Modern Living.* Kalamazoo, Mich.: Merchants Publishing Co., 1975.

Conover, C. A., and McElwee, E. W. *Selected Trees for Florida Homes.* Bulletin 182. Gainesville, Fla.: University of Florida Institute of Food and Agricultural Sciences, 1971.

Dickson, Felice. *Growing Food in South Florida.* Miami: Banyan Books, 1975.

Dworkin, Florence, and Dworkin, Stnaley. *The Apartment Gardener.* New York: Signet, 1974.

Fitch, Charles Marden. *The Complete Book of Houseplants.* New York: Hawthorne, 1972.

Florida Market Bulletin. A free classified advertising service that includes seeds and plants, published by the Florida Department of Agriculture and Consumer Services, Mayo Building, Tallahassee, Florida 32304.

Graf, Alfred. *Exotic House Plants.* 9th ed. East Rutherford, N.J.: Roehrs Co., 1975.

Graf, Alfred. *Exotic Plant Manual.* 3rd ed. East Rutherford, N.J.: Roehrs Co., 1974.

Graf, Alfred. *Exotica.* 8th ed. East Rutherford, N.J.: Roehrs Co., 1976.

Maxwell, Lewis, *Florida Flowers*. Tampa, Fla.: Lewis Maxwell, 1966.

Maxwell, Lewis. *Florida Fruit*. Tampa, Fla.: Lewis Maxwell, 1973.

Maxwell, Lewis. *Florida Insects*. Tampa, Fla.: Lewis Maxwell, 1974.

Maxwell, Lewis. *Florida Lawns and Gardens*. Tampa, Fla.: Lewis Maxwell, 1976.

Maxwell, Lewis. *Florida Plant Selector.* Tampa, Fla.: Lewis Maxwell, 1961.

Maxwell, Lewis. *Florida Vegetables*. Tampa, Fla.: Lewis Maxwell, 1974.

Menninger, Edwin A. *Flowering Trees of the World*. Great Neck, N.Y.: Hearthside, 1962.

Menninger, Edwin A. *Flowering Vines of the World*. Great Neck, N.Y.: Hearthside, 1970.

Menninger, Edwin A. *Seaside Plants of the World*. Great Neck, N.Y.: Hearthside, 1964.

Morton, Julia F. *Exotic Plants*. New York: Golden Press, 1971.

Morton, Julia F. *500 Plants of South Florida*. Miami: E. A. Seemann, 1975.

Morton, Julia F. *Orchids*. New York: Golden Press, 1970.

Morton, Julia F. *Wild Plants for Survival in South Florida*. Tampa, Fla.: Trend House, 1974.

Padilla, Victoria. *Bromeliads*. New York: Crown, 1973.

Perry, Mac. *Landscape Your Florida Home*. Miami: E. A. Seemann, 1972.

Ray, Richard M., ed. *House Plants Indoors/Outdoors*. San Francisco: Chevron Chemical Co., 1974.

Smiley, Nixon. *Florida Gardening Month by Month*. Rev. ed. Coral Gables, Fla.: University of Miami Press, 1970.

South Florida Orchid Society. *SFOS Culture Notes*. Questions and answers on orchid growing in South Florida. South Florida Orchid Society, Inc., 1900 S.W. Third Avenue, Miami, Florida 33129.

Sturrock. D. *Fruits for Southern Florida*. Stuart, Fla.: Southeastern Printing, 1959.

Watkins, John V. *Florida Landscape Plants, Native and Exotic*. Gainesville, Fla.: University of Florida Press, 1968.

Watkins, John V., and Wolfe, Herbert S. *Your Florida Garden*. Gainesville, Fla.: University of Florida Press, 1968.

Index

1. General Garden Care

2. Lawns

3. Shrubs, Foliage Plants, and Vines

4. Ornamental Trees

5. Palms

6. Fruit

112; oozing, 110; Pollock, 110-111, 112, 113; seedling, 111-112; sprouting, 110
avocado red mites, 101
bananas, 113-115
Barbados cherry, 93
bark, peeling, 102; splitting, 105-106, 109-110
bee, 112; leaf-cutting, 108
Benlate, 97, 98, 100
Benomyl, 97
birds, 126
black film *(see* mold, sooty)
black spotting *(see* anthracnose; scale)
blackfly, 102
bleach, 99
bloom spikes, on mango, 98; failure to set, 100-101
branch dieback, 96, 106
bulbul, 101
butterflies, swallowtail, 108
calamondin, 95, 107, 120
calcium, 98, 99
carambola, 93, 118, 119
Caribbean fruit fly, 98, 119-120
carissa, 123; *C. grandiflora,* 58, 59
caterpillars, 108
cherries, Barbados, 122; Surinam, 120
chicken manure, 117
chlordane, 115
chloride, 99
chlorosis, 102 *(see also* nutritional deficiencies)
Chrysophyllum spp., 95
citrus, 52, 73, dry fruit on, 102-103; fungi on, 102
copper, neutral, 95, 97-98, 102, 104, 105, 106, 109, 110, 119
County Cooperative Extension Service, 96
crown rot *(see* rot)
Cygon, 120
DeFend, 120
defoliation, 99, 100
dieback, 99, 106
drought, 96-97
Dylox, 120
eggfruit, 120
Epson salts, 115
Ferbam, 101, 118
fertilizing, of avocados, 111; of carambola, 118, of mangos, 98; over-, 95; of papayas, 117
figs, 120, 120-121
fish scraps, 117
flowers, dropping of, 108, 116
fly, fruit, 98, 102, 115, 117, 119-120

"foot rot" *(see* rot)
Formothion, 120
fruit, browning and blotching of, 100; drop, on mangos, 99, 100, on limes, 109; dry, 102-103; failure to, 106; greenish (oranges), 104; holes in, 101; rot, 107; splitting, 96-97, 104
fruit trees, ornamental, 93
fungicide, neutral copper, 95, 97-98
fungus, 95, 97-98, 106, 123
fuzz, red, 109
grafting, 96
grapefruit, 105, 106
greasy spot, 101, 109, 110
guava, 119-120; fruit flies on, 102
Hawaii Agricultural Experiment Station, 117
herbicides, 95
holes in fruit, 101
hornworms, 115
hurricane damage, 113
insects, 95, 98
iron, chelated, 94, 103, 121, 122
jaboticaba, 124
"jelly seed," 98
katydid, 108
kumquat, 93, 106-107, 120
leaves, black, spotty, *(see* mold, sooty) 101-102; burned looking, 100; chewed, 108-109; curled, 105; dropping, 101-102, 106, 109-110; small, distorted, 97; spotted *(see* spots); tip browning on, 95, 102
lemons, 109, 109-110
lichen, 97
lightning damage, 95
lime, 109; Key, 106, 107-108; Persian, 108-109
loquat, 118, 120
lychees, 94, 121-122
macadamia, 123
magnesium, 98, 99, 103, 115
Malathion, 99, 102, 105, 107, 108, 111, 120
Mammea americana, 95
Maneb, 118
manganese 104
mango, 95; booklet on, 96; failure to bloom, 100-101; fertilizing, 99; fruit drop on, 96, 99; fruit flies on, 98; fruit splitting on, 96-97; fruit rot on, 98; grafting, 96; holes in, 101; leaf drop on, 101; ripening time of, 96; scab on, 100; spraying, 97-98
Manilkara zapota, 95
melanose, 101
mildew, powdery, 99, 100, 115
minor element deficiency *(see* nutritional deficiency)

7. *Vegetables*

8. Annuals, Perrennials, and Bulbs

9. Orchids, Bromeliads, and Ferns

10. *Indoor and Terrace Plants*

11. *Propagation*